Metamaterials and Metasurfaces

Metamaterials and Metasurfaces

Editors

Naser Ojaroudi Parchin
Mohammad Ojaroudi
Raed A. Abd-Alhameed

MDPI • Basel • Beijing • Wuhan • Barcelona • Belgrade • Manchester • Tokyo • Cluj • Tianjin

MDPI

Editors
Naser Ojaroudi Parchin
School of Computing,
Engineering & the Built
Environment
Edinburgh Napier University
Edinburgh
United Kingdom

Mohammad Ojaroudi
COGNISCAN
Limoges
France

Raed A. Abd-Alhameed
School of Engineering and
Informatics
University Bradford
Bradford
United Kingdom

Editorial Office
MDPI
St. Alban-Anlage 66
4052 Basel, Switzerland

This is a reprint of articles from the Special Issue published online in the open access journal *Electronics* (ISSN 2079-9292) (available at: www.mdpi.com/journal/electronics/special_issues/metamaterials_metasurfaces).

For citation purposes, cite each article independently as indicated on the article page online and as indicated below:

LastName, A.A.; LastName, B.B.; LastName, C.C. Article Title. *Journal Name* **Year**, *Volume Number*, Page Range.

ISBN 978-3-0365-8035-7 (Hbk)
ISBN 978-3-0365-8034-0 (PDF)

Contents

About the Editors

Naser Ojaroudi Parchin

Naser Ojaroudi Parchin (SMIEEE) is currently an assistant professor (lecturer) at Edinburgh Napier University, Edinburgh, UK. He obtained his Ph.D. in Electrical Engineering from the University of Bradford, UK, where he was a postdoctoral research assistant at the Faculty of Engineering and Informatics and worked as a research fellow for the SATNEX V project, funded by the European Space Agency. From 2018 to 2020, Naser was a Marie Curie Research Fellow of the H2020-ITN-SECRET project funded by the EU Commission, targeting 5G mobile small cells. His research interests include phased arrays, MIMO systems, smartphone antennas, SAR/user impact, full-duplex diversity antennas, 5G antennas, implementable and biomedical sensors, RFID tag antennas, millimeter-wave/terahertz components, fractal structures, metamaterials and metasurfaces, PCB realization, Fabry resonators, EBG/FSS-inspired radiators, microwave filters, reconfigurable structures, and wireless propagation. He has over 12 years of research experience in antennas and microwave engineering. He has participated in several international IEEE conferences across the world, where he presented his articles through oral presentations. Dr. Naser is the author and co-author of several books/book chapters and more than 350 technical journals and conference papers. He is a member of the Institute of Electrical and Electronics Engineers (IEEE), the Marie Curie Alumni Association (MCAA), the European Association on Antennas and Propagation (EurAAP), and the Engineering Professors' Council (EPC). He is also an active reviewer for various high-ranking journals and publishers. He has been appointed as a Guest Editor and Topic Board Member for several Special Issues. He was included in the World's Top Scientists list in 2016, 2020, 2021, 2022, and 2023. His papers have more than 6600 citations with a 46 h-index, as reported by Google Scholar. Naser's score is higher than 95% of all RG members' scores.

Mohammad Ojaroudi

Mohammad Ojaroudi received his Ph.D. in electrical engineering from Shahid Beheshti University, Tehran, Iran. Over the past two decades, his research background has flourished at renowned academic institutions, where he actively took on the role of project manager in diverse research undertakings. The fruitful outcomes of these investigations have been disseminated through the publication of over 150 articles. Guided by an unwavering passion for comprehending consciousness and information processing within the brain, he has directed his focus toward pioneering advancements in neurotechnology. This pursuit led him to establish COGNISCAN in 2021, with an ambitious vision for harnessing the potential of electromagnetic waves in medical brain imaging. As the CEO of COGNISCAN, his primary responsibility lies in leading the executive team. Drawing upon his discerning judgment, he critically evaluates circumstances, establishes strategic objectives, and develops operational plans. He is also proficient in talent acquisition, nurturing a team of skilled professionals, and motivating and rewarding them to foster innovation and achieve tangible results.

Raed A. Abd-Alhameed

Raed A. Abd-Alhameed received his B.Sc. and M.Sc. degrees from the University of Basrah, Basrah, Iraq, in 1982 and 1985, respectively, and his Ph.D. degree from the University of Bradford, Bradford, U.K., in 1997, all in electrical engineering. He is currently a professor of electromagnetic and radiofrequency engineering. He has also been an investigator and co-investigator in several funded research projects. He has many years of research experience in the areas of radio frequency, signal processing, propagation, antennas, and electromagnetic computational techniques; he has published more than 500 academic journal articles and conference papers, and has coauthored three books and several book chapters. His research interests include computational methods and optimizations, wireless and mobile communications, sensor design, EMC, beam-steering antennas, energy-efficient PAs, and RF pre-distorter design applications. He is a Fellow of the Institution of Engineering and Technology, U.K., and of the Higher Education Academy, as well as being a registered Chartered Engineer in the U.K. He received the Business Innovation Award for his successful KTP with Pace and Datong's companies on the design and implementation of MIMO sensor systems and antenna array design for service localizations. He is the chair of several successful workshops on the topic of "Energy Efficient and Reconfigurable Transceivers: Approach Towards Energy Conservation and CO_2 Reduction" that address the biggest challenges for future wireless systems.

electronics

Editorial

Special Issue "Metamaterials and Metasurfaces"

Naser Ojaroudi Parchin [1,*], Mohammad Ojaroudi [2] and Raed A. Abd-Alhameed [3]

1 School of Computing, Engineering and the Built Environment, Edinburgh Napier University, Edinburgh EH10 5DT, UK
2 COGNISCAN, 1 Avenue d'Ester, CEDEX, 87280 Limoges, France; m.ojaroudi@cogniscan.fr
3 Faculty of Engineering and Informatics, University of Bradford, Bradford BD7 1DP, UK; r.a.a.abd@bradford.ac.uk
* Correspondence: n.ojaroudiparchin@napier.ac.uk

Metamaterials and metasurfaces have emerged as promising technologies in the field of antennas and wireless applications. They offer unprecedented control over electromagnetic waves, enabling the design of novel antenna structures with enhanced performance capabilities. By integrating metamaterial structures into antenna designs, it is possible to achieve electrically small antennas without sacrificing efficiency or bandwidth. Metasurfaces, on the other hand, are planar arrangements of subwavelength elements that manipulate the properties of incident electromagnetic waves. They are constructed by patterning a surface with precisely engineered meta-atoms, tailored to achieve specific functionalities. Metasurfaces can be used to control wavefronts, as well as the polarization and reflection properties of antennas. Metasurface antennas offer numerous advantages in wireless applications. They can achieve beamforming and steering capabilities without the need to use bulky and complex phased arrays in future wireless networks. Furthermore, metasurfaces can be used to create conformal and flexible antenna structures. This opens up new possibilities for the utilization of Internet of Things (IoT) applications and wearable devices. The use of metamaterials and metasurfaces can lead to significant advances in wireless communication systems, including improved signal quality, increased data rates, and seamless integration into various devices and environments.

The scope of this Special Issue encompasses a comprehensive exploration of metamaterials and metasurfaces, covering every facet of their design and construction. Moreover, it strives to spotlight captivating advances, prevailing trends, and recent accomplishments in this field. This Special Issue is a collection of 10 papers that are briefly explained in the following.

Lopato et al. [1] examine the impact of fabrication process uncertainties on the quality of terahertz metasurfaces, specifically focusing on how inaccuracies in metasurface fabrication affect resonances. They employ a numerical model to analyze the influence of uncertainties in the different geometric parameters obtained during the fabrication process, including layer deposition, photolithography, and etching processes, with respect to the resonance behavior of the designed metasurface. To validate their findings, the researchers verify the developed numerical model by applying it to a fabricated structure.

Wang et al. [2] investigate the challenge of anti-jamming matching reception for small-signal anti-jamming in the presence of intense electromagnetic jamming. The authors employ the Charnes–Cooper (CC) transform algorithm to identify the most favorable dynamic metamaterial antenna (DMA) array–element–codeword–state matrix, which maximizes the received signal-to-interference-plus-noise ratio (SINR). Through simulations, it is demonstrated that DMA offers notable advantages over traditional array antennas. Moreover, the utilization of DMA leads to reduced communication overhead, with a promising potential to upgrade existing wireless communication systems.

Liu et al. [3] delve into the exploration of negative group delay (NGD) metamaterials by utilizing split-ring resonators (SRRs). Their theoretical analysis involves the calculation of equivalent circuit parameters for two distinct types of SRRs. The measured results of the

Citation: Ojaroudi Parchin, N.; Ojaroudi, M.; Abd-Alhameed, R.A. Special Issue "Metamaterials and Metasurfaces". *Electronics* **2023**, *12*, 2420. https://doi.org/10.3390/electronics12112420

Received: 18 May 2023
Accepted: 23 May 2023
Published: 26 May 2023

prototypes align closely with the theoretical predictions and simulated outcomes. In addition, through simulations, it is demonstrated that the proposed metamaterials effectively reduce beam walk, showcasing their potential to serve as a solution in this context.

Zhang et al. [4] present the development of an ultra-broadband and angular-stable reflective linear into a cross-polarization converter utilizing a metasurface. The converter's unit cell is constructed using a slant end-loaded H-shaped resonator. Through simulations, is the authors reveal that the proposed design achieves a polarization conversion ratio exceeding 90% within the frequency range of 9.83–29.37 GHz, resulting in a relative bandwidth of 99.69% with high efficiency, ultra-broadband capability, and angular stability.

Amer et al. [5] present a polarization-insensitive broadband metamaterial absorber structure that exhibits wide-angle reception capabilities. The structure is based on square split-ring resonators (SSRRs) and incorporates lumped resistors. The proposed metamaterial absorber achieves absorption levels exceeding 90% over a wide frequency range from 1.89 GHz to 6.85 GHz, with a relative bandwidth of 113%.

Zhang et al. [6] demonstrate an inverse design framework for isomorphic metasurfaces utilizing representation learning. Through the use of autoencoders (AEs) with various architectures, the original high-dimensional space is effectively mapped onto a low-dimensional space with minimal information loss. It achieves a remarkable average accuracy of 94% on test sets, while also providing the design matrix within a matter of seconds, meaning that it significantly saves resources and time compared to traditional methods.

Voronov et al. [7] introduce new configurations of a magnetoinductive device that exhibits directional filter properties. Additionally, a new method is presented to enhance the device's filtering performance by compensating for multipath loss. The authors demonstrate techniques for constructing tunable devices that utilize toroidal ferrite-cored transformers in which experimental results confirm the agreement with the theoretical models.

Tavora de Albuquerque Silva et al. [8] introduce a novel unit cell design for electromagnetic bandgap (EBG) structures, utilizing a HoneyComb geometry (HCPBG). The design offers several advantages, including a reduced occupied area and flexible rejection band properties. Additionally, a strategy for the design of reconfigurable HCPBG filters is presented where the resonance frequency can be adjusted. The behavior and reconfiguration of the HCPBG filter are demonstrated through EM simulations.

Mitra et al. [9] employ transformation electromagnetics/optics (TE/TO) to realize a non-homogeneous, anisotropic material-embedded beam-steering technique without the need for phase control circuitry. The theoretical framework is supported by numerical simulations, validating the feasibility of the proposed approach. These innovative designs and methods have practical applications in various domains such as wireless communications, radar systems, beamforming, and beam steering.

Abdalrazak et al. [10] conduct a comprehensive literature review focusing on the squint phenomena in antenna arrays at mmWave frequencies. The study examines the challenges associated with the squint phenomena. The authors categorize the main effective solutions into beamforming techniques, antenna geometry modifications, and channel estimation algorithms. Additionally, another classification is explored, specifically one that leverages the beam squint phenomenon to improve channel localization and capacity. The literature review provides valuable insights into understanding and mitigating the challenges associated with squint phenomena in mmWave antenna arrays.

We would like to express our sincere appreciation and gratitude to all the authors who have made exceptional contributions to this journal. We would also like to extend our heartfelt thanks to the reviewers for their valuable comments and feedback, which have greatly improved the quality of the articles. Additionally, we would like to acknowledge the editorial board and the editorial office of *Electronics* for their support and guidance throughout the publication process. We hope that our readers will find the articles in this journal informative, insightful, and full of new and valuable information on Metamaterials and Metasurfaces for wireless communications.

Conflicts of Interest: The authors declare no conflict of interest.

References

1. Lopato, P.; Herbko, M.; Gora, P.; Mescheder, U.; Kovacs, A.; Filbert, A. Numerical Analysis of the Influence of Fabrication Process Uncertainty on Terahertz Metasurface Quality. *Electronics* **2023**, *12*, 2198. [CrossRef]
2. Wang, Y.; Jin, L.; Lou, Y.; Hao, Y. Small Signal Anti-Jamming Scheme Based on a DMA Linear Array under Strong Jamming. *Electronics* **2023**, *12*, 1389. [CrossRef]
3. Liu, Z.; Zhang, J.; Lei, X.; Gao, J.; Xu, Z.; Li, T. Negative Group Delay Metamaterials Based on Split-Ring Resonators and Their Application. *Electronics* **2023**, *12*, 1064. [CrossRef]
4. Zhang, B.; Zhu, C.; Zhang, R.; Yang, X.; Wang, Y.; Liu, X. Ultra-Broadband Angular-Stable Reflective Linear to Cross Polarization Converter. *Electronics* **2022**, *11*, 3487. [CrossRef]
5. Amer, A.A.G.; Sapuan, S.Z.; Alzahrani, A.; Nasimuddin, N.; Salem, A.A.; Ghoneim, S.S.M. Design and Analysis of Polarization-Independent, Wide-Angle, Broadband Metasurface Absorber Using Resistor-Loaded Split-Ring Resonators. *Electronics* **2022**, *11*, 1986. [CrossRef]
6. Zhang, J.; Yuan, J.; Li, C.; Li, B. An Inverse Design Framework for Isotropic Metasurfaces Based on Representation Learning. *Electronics* **2022**, *11*, 1844. [CrossRef]
7. Voronov, A.; Syms, R.R.A.; Sydoruk, O. High-Performance Magnetoinductive Directional Filters. *Electronics* **2022**, *11*, 845. [CrossRef]
8. Tavora de Albuquerque Silva, A.; Ferreira Dias, C.; Rodrigues de Lima, E.; Fraidenraich, G.; Medeiros de Almeida, L. A New Reconfigurable Filter Based on a Single Electromagnetic Bandgap Honey Comb Geometry Cell. *Electronics* **2021**, *10*, 2390. [CrossRef]
9. Mitra, D.; Dev, S.; Allen, M.S.; Allen, J.W.; Braaten, B.D. Coordinate Transformations-Based Antenna Elements Embedded in a Metamaterial Shell with Scanning Capabilities. *Electronics* **2021**, *10*, 1081. [CrossRef]
10. Abdalrazak, M.Q.; Majeed, A.H.; Abd-Alhameed, R.A. A Critical Examination of the Beam-Squinting Effect in Broadband Mobile Communication: Review Paper. *Electronics* **2023**, *12*, 400. [CrossRef]

electronics

Article

Numerical Analysis of the Influence of Fabrication Process Uncertainty on Terahertz Metasurface Quality

Przemyslaw Lopato [1,*], Michal Herbko [1,*], Paulina Gora [1], Ulrich Mescheder [2], Andras Kovacs [2] and Alexander Filbert [2]

[1] Center for Electromagnetic Fields Engineering and High-Frequency Techniques, Faculty of Electrical Engineering, West Pomeranian University of Technology in Szczecin, ul. Sikorskiego 37, 70-313 Szczecin, Poland

[2] Institute for Microsystems Technology (iMST), Faculty of Mechanical & Medical Engineering, Robert Gerwig-Platz 1, 78120 Furtwangen im Schwarzwald, Germany

* Correspondence: plopato@zut.edu.pl (P.L.); michal.herbko@zut.edu.pl (M.H.)

Abstract: The purpose of this paper was to investigate the influence of fabrication process uncertainty on terahertz metasurface quality. The focus was on the effect of metasurface fabrication inaccuracy on resonances. To the best of our knowledge, this is the first paper to study the effect of the metasurface fabrication process on its resonant frequency. The terahertz split ring resonator-based metasurface is under consideration. Using a numerical model, the influence of the uncertainty of various geometrical parameters obtained during the fabrication process (mainly layer deposition, photolithography, and etching processes) is analyzed according to the resonance of the designed metasurface. The influence of the following parameters causes a shift of resonant frequencies of the considered metasurface: etching deviation e, metallization thickness t_{Al} and SiO_2 layer thickness t_{SiO2}. The quality of the metasurface affected by the variations of obtained geometrical parameters was determined by the deviation of resonant frequency Δf_r. The developed numerical model was verified by THz-TDS (terahertz time-domain spectroscopy) measurements of the fabricated structure.

Keywords: metasurfaces; terahertz devices; electromagnetic waves; finite element model

Citation: Lopato, P.; Herbko, M.; Gora, P.; Mescheder, U.; Kovacs, A.; Filbert, A. Numerical Analysis of the Influence of Fabrication Process Uncertainty on Terahertz Metasurface Quality. *Electronics* **2023**, *12*, 2198. https://doi.org/10.3390/electronics12102198

Academic Editors: Raed A. Abd-Alhameed, Naser Ojaroudi Parchin, Mohammad Ojaroudi and Giovanni Crupi

Received: 2 March 2023
Revised: 5 May 2023
Accepted: 10 May 2023
Published: 12 May 2023

1. Introduction

A metamaterial (MM) is an artificial material/device consisting of an array of sub-wavelength-size structural elements (shown in Figure 1) that can take on the properties of non-naturally occurring materials [1]. It can be more specifically described as a ε-negative (ENG) or μ-negative (MNG) material of either negative permittivity or permeability, respectively, a double-negative (DNG) when both of these quantities are simultaneously negative in a particular frequency range, and ε-near-zero (ENZ) or a μ-near-zero (MNZ) material with small, close-to-zero values for these quantities (this might be the case when the materials are close to their electric or magnetic plasma frequency) [2]. This creates new opportunities to interact with electromagnetic waves (from microwave band to visible light) and mechanical (ultrasonic) waves. In this paper, we will focus on electromagnetic waves in the terahertz frequency band. In this case, various field transformations of arbitrary incident fields can be achieved. Planar versions of MM-metasurfaces (MS) are much easier to fabricate (in comparison to 3D structures) and can potentially have many applications. The first of the areas in which metasurfaces can be used is telecommunications. Among other things, it can be used to improve the efficiency of antennas by increasing directivity, increasing power gain, increasing selectivity, or by multiplexing signals [3–6]. Other peripheral areas for metasurfaces are selective absorbing structures [7,8], polarizers [9,10], surface wave control devices [11,12], wave front manipulation devices [13,14] and imaging systems (for lensing electromagnetic waves) [15,16]. Particularly noteworthy are works showing plasmonic metasurfaces used as perfect absorbers and pulse controllers [17,18].

Other peripheral areas for metasurfaces are measurement systems. Among the most frequently mentioned are applications for evaluating biological samples, where it is most often indicated in the detection of biomolecules, microorganisms, cancer, or tumors [19]. Other applications include testing chemicals and measuring strain [20,21]. Exemplary metasurface structures are presented in Figure 1.

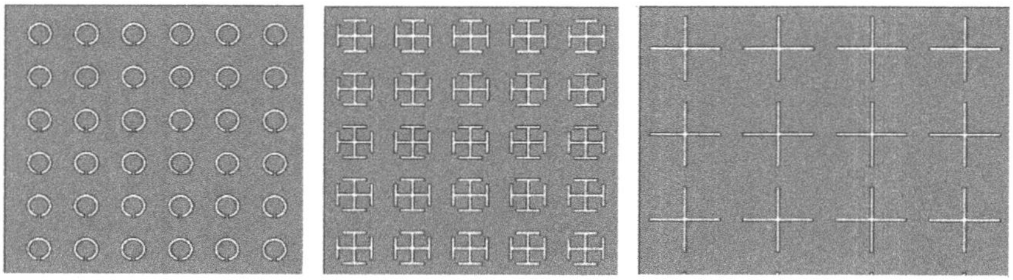

Figure 1. Exemplary metasurface structures.

Characterization of the fabrication process uncertainty is a key question for quality control. So far, the effect of changing the geometry of the metasurface created by inaccuracies in the fabrication process has been studied for different dimensions of rectangular structural elements in the work [22]. In our work, we investigated the effect of the fabrication process on the operation of SRR-based meta-surfaces; it is a more complicated structure to analyze because as a result of the process, some dimensions increase (gap width and distances between elements) and some decrease (line width and size of the square loop) as a result of over-etching. In our work, the effect of not only changing the geometry as a result of the etching process but also of oxidation and metallization was studied. The paper [22] studied the effect of the fabrication process on the metasurface in a hologram application, where the effect of fabrication on the phase of the electromagnetic wave was examined. In this paper, the effect of the fabrication process on resonances was studied, as we are interested in sensor applications. To the best of our knowledge, this is the first paper to study the effect of the fabrication process of a metasurface on its resonant frequency.

The fabrication of the MS in the terahertz range is carried out using a silicon-based microtechnology process. Although this technology is precise, there are some deviations from the planned geometric dimensions or thicknesses of individual layers. In this paper, the split-ring Resonator (SRR)-based metasurface was designed for the resonant frequency of $f_r = 1$ THz, and the influence of fabrication process uncertainty on terahertz metasurface quality was characterized using numerical analysis. In this article, the square version of the SRR was chosen. The quality of the metasurface affected by the received variation geometrical parameters was determined by the resonant frequency shift Δf_r. The increase in the metallization thickness t_{Al} as well as the SiO_2 layer thickness t_{SiO2} causes an increase in the resonant frequencies of the considered metasurface, while etching deviation e has the opposite effect. The ±20% parameter variation range was assumed. The resulting frequency deviations potentially causing a degradation of metasurface performance in some applications were examined. Thanks to the use of a validated numerical model, it was possible to estimate the deviations of electromagnetic parameters of the designed structures resulting from the inaccuracy of microfabrication. This problem becomes particularly important in the higher bands of the spectrum—infrared and visible light—where (as the electromagnetic wavelength decreases) the inaccuracies in the manufacturing process increase significantly. The developed numerical model was verified by THz-TDS measurements of the fabricated structure.

2. THz Metasurface Design and Fabrication

The arbitrary selected split-ring resonator metasurface was designed with the following assumptions: a primary resonant frequency of $f_{r1} = 1$ THz, and a secondary resonant frequency of $f_{r2} < 3$ THz. Moreover, we assumed that the magnitude of transmission coefficient obtained within simulations $|S_{21}| < -40$ dB for both resonances. Additionally, there were some restrictions according to the planned fabrication process that will be described later in this paper. The design (determination of dimensions) was based on a simple optimization procedure. In the first step based on the literature review, an initial set of geometrical parameters was developed. Relationships from the paper [23] were used to develop the initial design. Then, this model was implemented and simulated in the COMSOL Multiphysics environment. In the third step, the geometrical parameters describing the structural element were tuned to meet the assumed electromagnetic parameters (f_{r1}, f_{r2}, and $|S_{21}|$). The obtained dimensions are presented in Table 1 and the geometry of the squared SRR structural element with the description of dimensions is shown in Figure 2. The structure was developed on a 520 µm thick silicon (Si) wafer. First, the wafer was covered with a thermal oxide layer (SiO_2) (on both sides) with a thickness of 170 nm. The front side of the wafer was coated with thermally evaporated aluminum. The wafer was coated with an AZ 1518 photoresist layer and exposed with the designed terahertz filter mask. The aluminum layer was wet-chemically structured with a phosphoric acid etching mixture, and the photoresist layer was removed in an oxygen plasma. The individual chips for the measurement were separated with a diamond saw. The flowchart of the manufacturing process is shown in Figure 3. The obtained line width was 2.5 µm, while the capacitive gap was 2.5 µm and the length of the side of the square loop was 21.7 µm.

Table 1. Dimensions of designed SRR metasurface.

Dimensions	[µm]
Periodicity, p	24.2
Square loop size, a	21.7
Line width, w	2.5
Gap, g	2.5
Aluminum thickness, t_{Al}	0.5
Silicon dioxide thickness, t_{SiO2}	0.17
Silicon thickness, t_{Si}	520

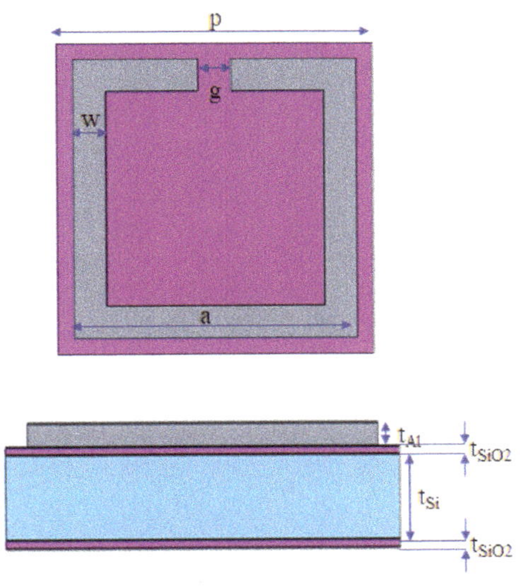

Figure 2. Dimensions of SRR metasurface on the silicon wafer with a top view (**top**) and cross-section (**bottom**).

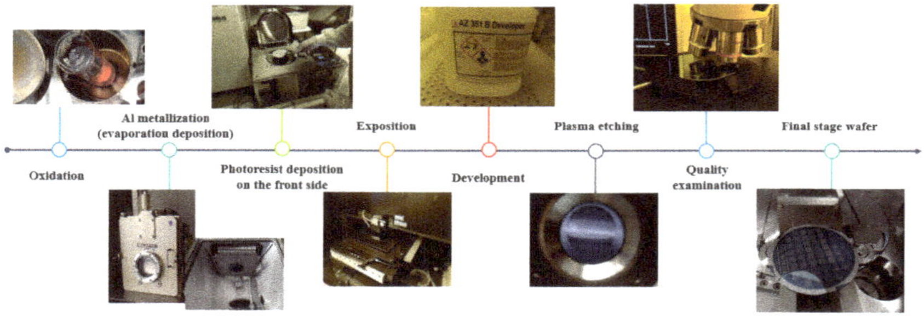

Figure 3. Flowchart of the manufacturing process.

The design and further analyses were carried out using a finite element method (FEM) based numerical model. All the computations were made in the COMSOL Multiphysics 6.0 software. The geometry of the model is shown in Figure 4. The following values of dielectric properties were assumed for the materials. The refractive index of silicon is n_{Si} = 3.48 (permittivity is ε_{rSi} = 12.11) [24]. The real and imaginary parts of the complex permittivity of silicon dioxide for the terahertz range are ε'_{rSiO2} = 4.45 and ε''_{rSiO2} = 0.03, respectively [25]. The Drude model was used to describe the value of the aluminum refractive index n_{Al} [26]:

$$n_{Al} = \sqrt{\varepsilon'_{rAl} + i\varepsilon''_{rAl}} = \sqrt{1 - \frac{\omega_p^2}{\omega^2 + \gamma^2} + i\frac{\omega_p^2\gamma}{\omega_p^3 + \gamma^2\omega}} \tag{1}$$

where ω_p is the plasma frequency of the conductor (depending on the dielectric constant in vacuum, the density of free electrons, electron mass, and charge), ω is the angular frequency of the incidence wave, and γ is the damping coefficient of the conductor. The plasma frequency for aluminum is $\omega_p = 2.123 \times 10^{16}$ rad/s, while the damping coefficient is $\gamma = 1.975 \times 10^{11}$/s.

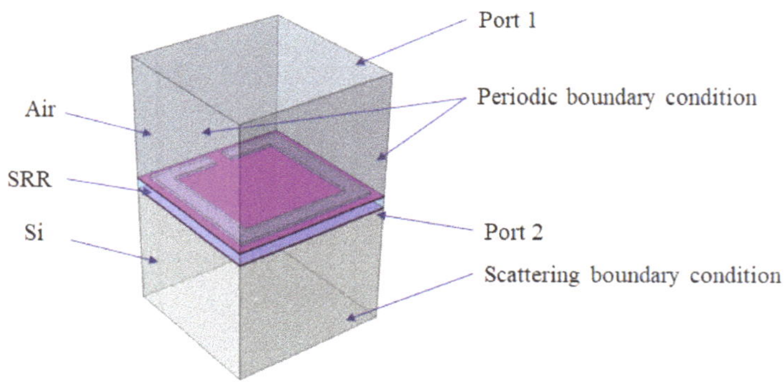

Figure 4. Numerical model of SRR metasurface on the silicon wafer (the thickness of the Si substrate not in scale).

The wave excitation and scattering matrix parameters calculation were enabled through ports (excitation and S_{11} calculation on port 1, S_{21} calculation on port 1 and 2). Port 1 is placed at some distance over the MS element separated by an air layer, while port 2 is placed on the end of the metasurface stack. The scattering boundary condition was set on the bottom boundary of the computational model in order to fully absorb the outgoing wave and omit its influence on the results including the Fabry–Perot effect. However, this effect can also occur in the case of the Si substrate. For this reason, the solution proposed in paper [27] was utilized: the Si wafer consisting of metasurface (not in scale in Figure 4) was backed with an Si layer (not an air layer such as in ordinary operating conditions), and the receiving port was placed on the back surface of the Si wafer. This causes a lack of reflection on the Si wafer's back layer boundary. Thus, this configuration makes it possible to eliminate the Fabry–Perot effect while conserving the original response of the metasurface. The Fabry–Perot effect strongly affects the results obtained in the terahertz frequency range (multiple, relatively high amplitude ripples obtained in spectrum); thus, in the case of frequency domain computations, some sophisticated methods for its reduction should be applied.

In this high-frequency electromagnetic problem the following frequency domain equation is solved [28,29]:

$$\nabla \times \mu_{\mathrm{r}}^{-1}(\nabla \times E) - k_0^2\left(\varepsilon_{\mathrm{r}} - \frac{j\sigma_e}{\omega\varepsilon_0}\right)E = 0 \tag{2}$$

where: E—electric field, μ_{r}—relative permeability, ε_0—permittivity of vacuum, ε_{r}—relative permittivity, σ_e—electric conductivity, k_0—wave number, ω—angular frequency. The exemplary mesh used during numerical calculations consists of 106,029 domain elements, 17,278 boundary elements, and 1602 edge elements.

To reduce the computational complexity of the developed model (instead of the simulation of a large matrix of structural elements) just a single-element analysis was performed. This simplification was based on the application of Floquet periodic boundary conditions on the side boundaries. Such simplification of the model is acceptable, as will be shown later in this paper, through the comparison of modeled and measured resonant frequencies and S_{21} reflection coefficient frequency responses.

To investigate the influence of the fabrication process and material properties on the quality of the designed SRR structure, the following parameters were analyzed: etching deviation e, metallization thickness t_{Al}, and SiO_2 layer thickness t_{SiO2}. The quality of the metasurface affected by the above parameters was determined by the deviation of resonant frequency Δf_{r}:

$$\Delta f_{\mathrm{r}} = f_{\mathrm{rD}} - f_{\mathrm{rB}} \tag{3}$$

where f_{rD}—resonant frequency of the structure with deviated parameters, f_{rB}—resonant frequency of the structure with a base set of parameters (shown in Table 1).

The influence of each of the above parameters on the behavior of the metasurface was determined using a developed finite element method (FEM)-based numerical model.

3. Results

The developed numerical model was utilized to perform the considered analyses. The exemplary results of the simulations are presented in Figure 5; the frequency response of transmission $|S_{21}|$ and the reflection $|S_{11}|$ scattering matrix parameters were presented for the dimensions given in Table 1. One can observe that there are two resonances: 1 THz and 2.77 THz. The distribution of the electric field on the surface of MS is shown for the first resonance (frequency of 1 THz). As was expected, there is an enhancement of the electric field intensity in the vicinity of the capacitive gap. The presented results show that the designed metasurface meets the requirements assumed in the design process. This positive verification allowed us to perform the fabrication process based on the dimensions presented in Table 1. The photo of the fabricated structure is shown in Figure 6.

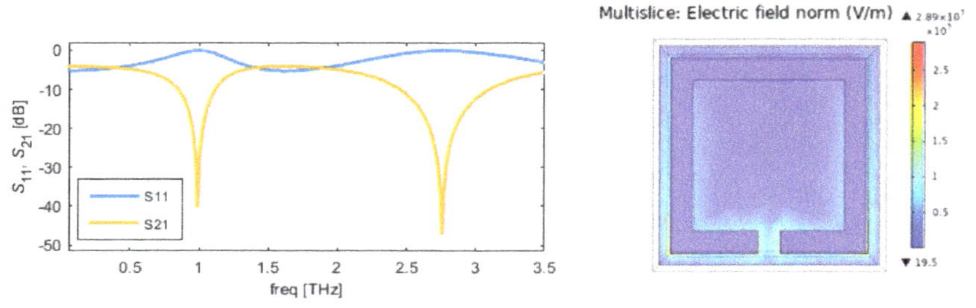

Figure 5. Results of FEM numerical simulations: frequency response (**left**) and electric field distribution on the surface of SRR (**right**).

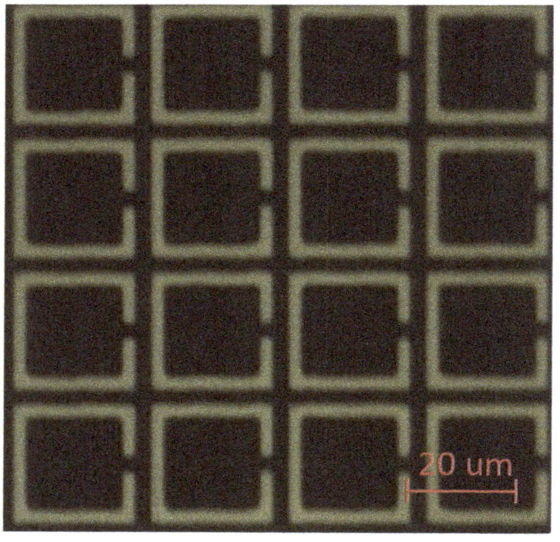

Figure 6. Photo of manufactured SRR metasurface on the silicon wafer (light area—Al metallization; dark area—lack of metallization, SiO_2).

The fabricated structure was experimentally evaluated to verify the results of the design process, as well as the numerical model and the applied assumptions (such as frequency domain computation, Floquet periodic boundary conditions, plane wave excitation, dispersive modeling of the metallization layer, etc.). The measurements were performed using the THz-TDS (terahertz time domain spectroscopy) system TeraFlash smart of Toptica, Germany, shown in Figure 7. A transmission setup was utilized, whereby the transmission Tx and receiver Rx heads are on opposite sides of the examined device/structure under test (DUT). The electromagnetic pulse with a terahertz frequency content is generated by a photoconductive antenna and focused on the evaluated metasurface structure using a set of curved reflectors. After interaction with the DUT, the modified pulse defocuses and propagates through another set of curved reflectors and is passed to the receiving photoconductive antenna. The focusing spot size (the area of incidence of the terahertz beam) is less than 1 mm diameter. The results of the measurements are presented in Figure 8. The presented time domain pulses were obtained for two scenarios: the Si wafer without metasurface (reference signal), and the Si wafer containing an MS. As the frequency domain data are of more importance in the considered analysis (resonant frequency analysis), a transformation of the data from the time to frequency domain was obtained using Fourier transform. In the first step, the measurement without DUT was made (air transmission spectrum) in order to verify the transmission band. In the obtained frequency response, the usable signal (above the noise level) can be observed for the frequency band below 2.5 THz. This means that the second resonance will not be possible to measure for this configuration (the frequency responses presented in Figure 8 obtained for the Si wafer also confirm a maximum usable frequency of 2.5 THz). Within the transmission band, there are multiple absorption peaks visible, caused by the absorption of certain frequencies by water, nitrogen, and oxygen molecules. In the next step, the measurement of the Si wafer without metallization was made to obtain the reference signal. Finally, the measurement of full DUT (the Si wafer with a metallization metasurface) was performed. One can observe the decrease in the transmission spectrum in the vicinity of a frequency of 1 THz, which is caused by the resonant behavior of MS. After de-embedding, the metasurface transmission spectrum consists of just one (primary) resonance for the desired frequency of 1 THz. The resonant frequency obtained in the simulations was f_{r1Sim} = 1.0023 THz, and in the measurements was f_{r1Meas} = 1.040 THz. Good agreement between the simulations and measurement results was proved by an absolute difference of 37.7 GHz and a relative difference of 3.76%. The measurement results are consistent with the resonance frequency fr = 1 THz assumed in the design process and is within the deviation of +46.47 GHz/−42.83 GHz determined (in a later part of this work) by numerical analysis. As indicated in the article, due to the limited bandwidth of the measurement system used, experimental verification in the case of the second resonance was not possible.

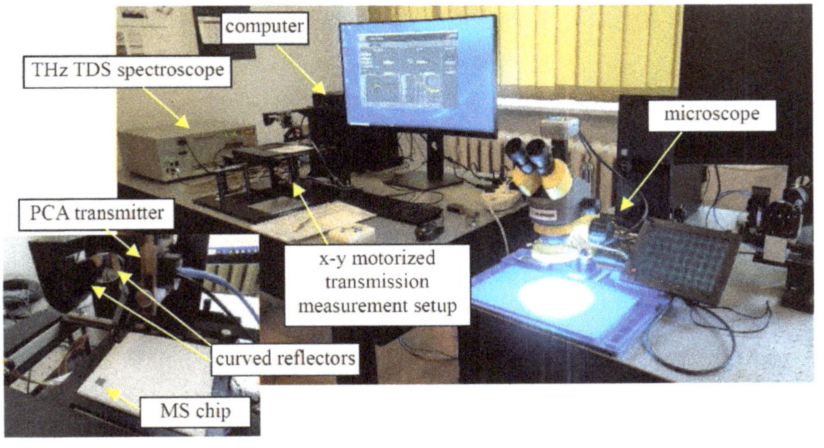

Figure 7. Photo of THz TDS measuring system.

After positive verification of the developed numerical model and all the assumptions made, analysis of the influence of selected geometrical parameters was performed. These parameters (etching deviation *e*, metallization thickness t_{Al}, and SiO_2 layer thickness t_{SiO2}) were selected as the most susceptible to possible deviations during the production process. The values of the t_{Al} and t_{SiO2} parameters presented in Table 1 were taken as reference values, and during the analysis, these values were changed in the range of ±20% of the reference value: $e = -20 \ldots +20\%$, $t_{Al} = 0.400 \ldots 0.600$ μm and $t_{SiO2} = 0.136 \ldots 0.204$ μm. The etching deviation parameter is a more complicated value, consisting of the line width *w*, which was changed from the range of ±20%, while simultaneously changing the gap *g* and the square loop size *a*. The considered values of *w*, *g*, and *a* for the different values of the etching deviation parameter *e* are shown in Table 2.

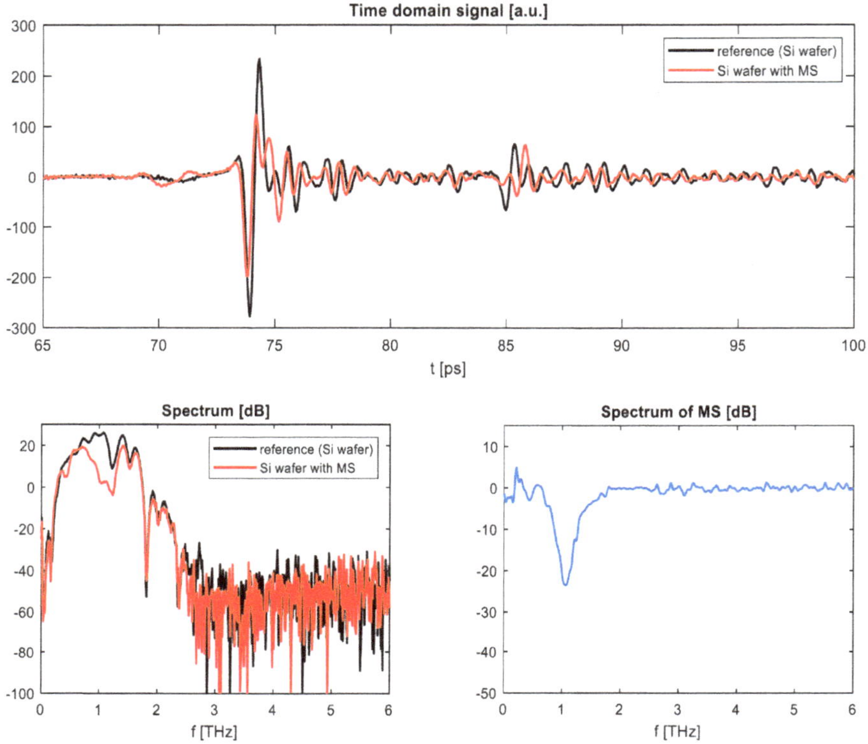

Figure 8. Results of measurements using THz TDS spectroscope. Results were obtained through the Fourier transform of time domain data.

Table 2. List of parameter sets for individual etching deviation parameter values *e*.

Etching Deviation e [%]	Line Width w [μm]	Gap g [μm]	Square Loop Size a [μm]
−20	2	3	21.2
−15	2.125	2.875	21.325
−10	2.25	2.75	21.45
−5	2.375	2.625	21.575
0	2.5	2.5	21.7
5	2.625	2.375	21.825
10	2.75	2.25	21.95
15	2.875	2.125	22.075
20	3	2	22.2

An exemplary result of such analysis—the shift of the resonant curve caused only by the change in the metallization layer thickness t_{Al}—is presented in Figure 9. One can observe the increase in resonant frequency f_{r1} while the metallization thickness t_{Al} increases. This relationship is presented for both resonances and all three considered parameters in Figure 10. In all cases we can observe a quasi-linear relationship when considering ±20% variation ranges.

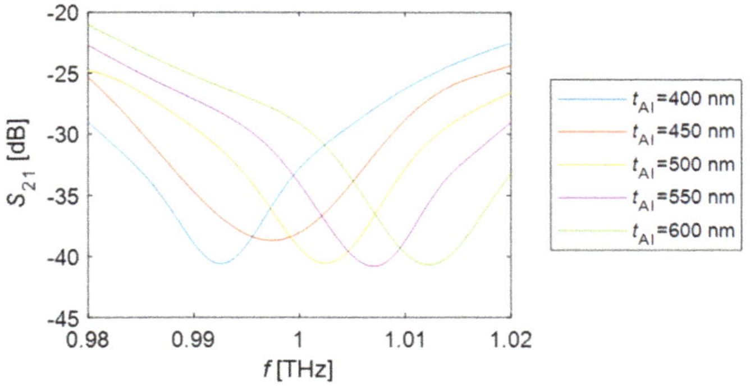

Figure 9. Results of FEM numerical simulations: shift of resonant curve caused by the thickness of the metallization layer t_{Al}.

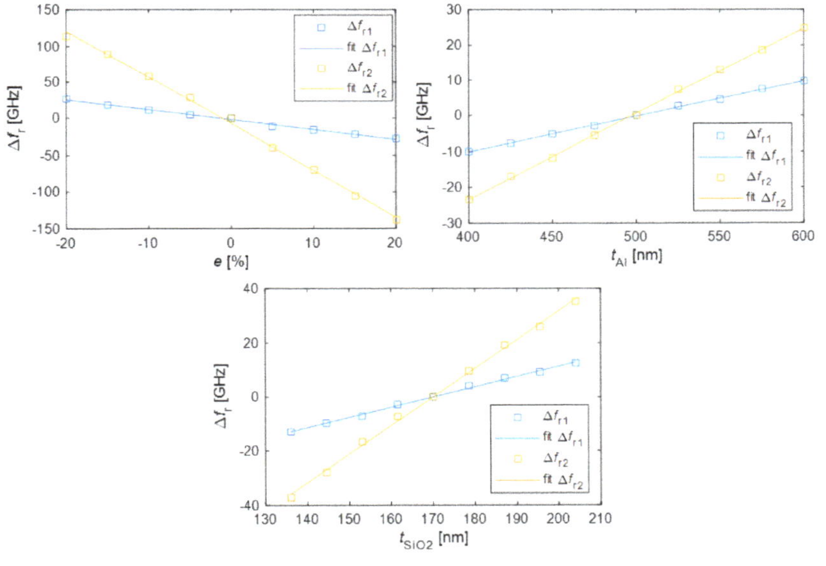

Figure 10. Numerically calculated influence of various geometrical parameters (e, t_{Al}, and t_{SiO2}) on the shift of resonant frequencies Δf_{r1} and Δf_{r2}: etching e (**left**), aluminum metallization thickness t_{Al} (**right**), and silicon dioxide layer thickness t_{SiO2} (**bottom**).

The influence of the analyzed parameters can be considered based on an equivalent circuit—for SRR—schematically modeled as a parallel resonant circuit with a resonance frequency f_{rSCH} [30]:

$$f_{rSCH} = \frac{1}{2\pi\sqrt{LC}} \tag{4}$$

where f_{rSCH}—resonant frequency of equivalent circuit, L—equivalent inductance, C—equivalent capacitance.

Increasing the thickness of the metallization increases the conductance but decreases the inductance of the wire constituting the resonant ring. Reducing the inductance in the equivalent circuit increases the resonant frequency.

The capacitance of the SRR structure depends to the greatest extent on the capacitance of the gap, which can be approximated by a flat capacitor, in which the electric field is partially distributed in air and partially in a silicon wafer, in particular the SiO_2 layer on the surface. Therefore, to determine the capacitance, it is necessary to use the equivalent permittivity (depending on the depth of the SiO_2 layer). In the case of a thin SiO_2 layer, part of the electric field exists in the Si (high permeability) layer. The increase in the thickness of the SiO_2 layer causes the electric field to be distributed to a greater extent in the material characterized by lower permittivity, so as a consequence the resultant capacitance decreases. A smaller value of the gap capacitance increases the resonant frequency.

According to Table 2, an increase in the etching deviation factor causes an increase in linewidth and square loop size and a decrease in capacitive gap size. An increasing linewidth causes a decrease in the inductance and, as a result, an increase in resonant frequency. A reduction in gap size causes a higher capacitance and consequently a lower resonant frequency. If the square loop size increases, the inductance increases too, thus the resonant frequency decreases. Finally, in the case of the etching deviation factor increasing, the resulting resonance moves to lower values.

The comparison of the influence of considered parameters on the maximum shift of resulting resonant frequencies is presented in Table 3. The comparison was divided into two stages: first, each of the considered parameters was analyzed separately, then two worst-case scenarios were verified:

- first worst case: the lowest considered etching deviation e, and the highest considered metallization layer thickness t_{Al}, and silicon dioxide thickness t_{SiO2}—causing an increase in resonant frequencies;
- second worst case: the highest considered etching deviation e, and the lowest considered metallization layer thickness t_{Al} and silicon dioxide thickness t_{SiO2}—causing a decrease in resonant frequencies;

Table 3. Comparison of various parameters' influences on the maximum shift of resulting resonant frequencies.

Uncertainty Source	Δf_{r1} [GHz]	Δf_{r2} [GHz]
Etching deviation, $e^{®} - 20\%$	25.39	120.41
Etching deviation, $e^{®} + 20\%$	−28.33	−134.70
Aluminum thickness, $t_{Al}{}^{®} - 20\%$	−10.02	−23.48
Aluminum thickness, $t_{Al}{}^{®} + 20\%$	9.82	24.80
Silicon dioxide thickness, $t_{SiO2}{}^{®} - 20\%$	−12.76	−35.91
Silicon dioxide thickness, $t_{SiO2}{}^{®} + 20\%$	12.98	36.02
First worst case (e) – 20%, (t_{Al} + t_{SiO2}) + 20%	46.47	170.61
Second worst case (e) + 20%, (t_{Al} + t_{SiO2}) − 20%	−42.83	−192.12

This was possible because of similar and monotonic characteristics of the relationships shown in Figure 10.

In the case of only one parameter variation, the biggest influence was caused by the line etching deviation e, while the smallest one was by aluminum thickness t_{Al}. The frequency shifts, as expected, were 2–5 times higher for the second resonance. Finally, in the case of the worst-case scenario (all the considered parameters deviations were taken into account), the maximum change in f_{r1} was close to 46 GHz and the change in f_{r2} was 192 GHz. This caused the following relative deviations of resonant frequencies: 4.6% for f_{r1} and 6.9% for f_{r2}. The shifts of resonant curves caused by the cumulative change in etching deviation e, aluminum layer thickness t_{Al}, and silicon dioxide thickness t_{SiO2} are presented in Figure 11.

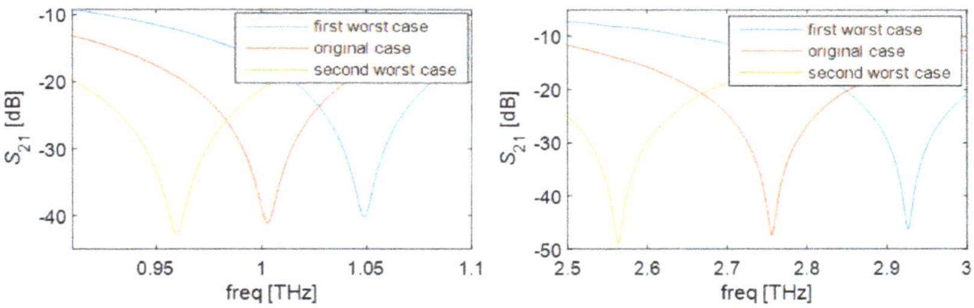

Figure 11. Results of FEM numerical simulations: shift of resonant curve caused by the cumulative change in etching deviation *e*, aluminum layer thickness t_{Al} and silicon dioxide thickness t_{SiO2}—worst cases.

4. Conclusions

In this paper, the design and fabrication of split-ring-resonator-based metasurfaces were performed and a further analysis of the fabrication process uncertainty on resulting MS parameters was numerically conducted. The process of terahertz metasurface fabrication is relatively complicated (lots of parameters need to be precisely controlled), and may be the source of some deviations from the planned (designed) dimensions resulting in shifts of assumed electromagnetic parameters. The sources of probable inaccuracies in the fabrication process were identified (etching process deviation *e*, aluminum metallization thickness t_{Al}, and silicon dioxide layer thickness t_{SiO2} variations) and their influence was analyzed using the numerical model. The commonly utilized metasurface structure of the square split-ring resonator was chosen and designed for operation in the terahertz frequency range. In this case, not the full matrix of structural elements but a single-element numerical analysis (with periodic boundary conditions) was utilized. This assumption regarding the periodicity enables the examination of various fabrication aspects, as long as there is no asymmetry in the metasurface's geometry.

It was assumed that the geometrical parameters (e, t_{Al}, and t_{SiO2}) were modified in the range of ±20%. All the evaluated sources of geometric deviation caused a linear change in both resonant frequencies. This caused the following relative deviations of resonant frequencies: 4.6% for f_{r1} and 6.9% for f_{r2}. Such frequency deviations can cause significant degradation of the MS performance in some applications. Depending on the application of a particular metasurface, other parameters (than resonant frequency) should be considered for evaluation; e.g., in the case of reconfigurable structures (flexible or MEMS devices), tunability range can be evaluated.

Author Contributions: Conceptualization, P.L., M.H., U.M. and A.K.; methodology, P.L., M.H., U.M. and A.K.; process development and fabrication, U.M., A.K. and A.F.; numerical modelling, M.H., P.G. and P.L.; measurements, P.L. and M.H.; writing—original draft preparation, P.L, M.H. and A.K.; writing—review and editing, P.L. and A.K.; project administration, P.L. and U.M.; funding acquisition, P.L. and U.M. All authors have read and agreed to the published version of the manuscript.

Funding: This work was conducted within the project "RE-TERA- Reconfigurable terahertz devices for EM waves manipulation and sensing applications", co-financed by the Polish National Agency for Academic Exchange (NAWA, Poland) and German Academic Exchange Service (DAAD, Germany) under the grant no. PPN/BDE/2021/1/00012/U/00001/ID: 57602825.

Acknowledgments: We would like to thank Xenia Seng from the Institute of Microsystems Technology, Furtwangen University, for carrying out the microfabrication process.

Conflicts of Interest: The authors declare no conflict of interest.

References

1. Valipour, A.; Kargozarfard, M.H.; Rakhsi, M.; Yaghootian, A.; Sedighi, H.M. Metamaterials and their applications: An overview. *Proc. Inst. Mech. Eng. Part L J. Mater. Des. Appl.* **2021**, *236*, 2171–2210. [CrossRef]
2. Engheta, N.; Alu, A.; Silveirinha, M.G.; Salandrino, A.; Li, J. DNG, SNG, ENZ and MNZ Metamaterials and Their Potential Applications. In Proceedings of the MELECON 2006–2006 IEEE Mediterranean Electrotechnical Conference, Malaga, Spain, 16–19 May 2006. [CrossRef]
3. Nie, S.; Akyildiz, I.F. Metasurfaces for multiplexed communication. *Nat. Electron.* **2021**, *4*, 177–178. [CrossRef]
4. Ali, A.; Khalily, M.; Brown, T.W.; Tafazolli, R. Metasurface-based THz Reflectarray Antenna with Vortex Multiplexing and Beam-steering Capabilities for Future Wireless Communications. *iScience* **2022**, *25*, 104725. [CrossRef] [PubMed]
5. Boyarsky, M.; Sleasman, T.; Imani, M.F.; Gollub, J.N.; Smith, D.R. Electronically steered metasurface antenna. *Sci. Rep.* **2021**, *11*, 4693. [CrossRef]
6. Huang, R.-Z.; Zhang, J.-W.; Zhang, C. Dual-Band Circularly Polarized Hybrid Dielectric Resonator Antenna for 5G Millimeter-Wave Applications. *Electronics* **2022**, *11*, 1761. [CrossRef]
7. Alaee, R.; Albooyeh, M.; Rockstuhl, C. Theory of metasurface based perfect absorber. *J. Phys. D* **2017**, *50*, 503002. [CrossRef]
8. Zhou, Q.; Ma, W.; Wu, T.; Li, Y.; Qiu, Q.; Duan, J.; Li, J.; Jiang, L.; Zhou, W.; Gao, Y.; et al. Metasurface Terahertz Perfect Absorber with Strong Multi-Frequency Selectivity. *ACS Omega* **2022**, *7*, 36712–36727. [CrossRef]
9. Kurosawa, H.; Choi, B.; Sugimoto, Y.; Iwanga, M. High-performance metasurface polarizers with extinction ratios exceeding 1200, High-performance metasurface polarizers with ratios exceeding 12000. *Opt. Express* **2017**, *25*, 4446–4455. [CrossRef]
10. Wang, S.; Deng, Z.L.; Wang, Y.; Zhou, Q.; Wang, X.; Cao, Y.; Guan, B.O.; Xiao, S.; Li, X. Arbitrary polarization conversion dichroism metasurfaces for all-in-one full Poincaré sphere polarizers. *Light Sci. Appl.* **2021**, *10*, 24. [CrossRef]
11. Martini, E.; Mencagli, M., Jr.; Maci, S. Metasurface transformation for surface wave control. *Phil. Trans. R. Soc. A* **2015**, *373*, 20140355. [CrossRef]
12. Kiani, M.; Momeni, A.; Tayarani, M.; Ding, C. Spatial wave control using a self-biased nonlinear metasurface at microwave frequencies. *Opt. Express* **2020**, *28*, 35128–35142. [CrossRef]
13. Li, J.; Yue, Z.; Li, J.; Zheng, C.; Zhang, Y.; Yao, J. Ultra-narrowband terahertz circular dichroism driven by planar metasurface supporting chiral quasi bound states in continuum. *Opt. Laser Technol.* **2023**, *161*, 109173. [CrossRef]
14. Li, J.; Yue, Z.; Li, J.; Zheng, C.; Wang, S.; Li, M.; Zhang, Y.; Zhang, Y.; Yao, J. Diverse terahertz wavefront manipulations empowered by the spatially interleaved metasurfaces. *Sci. China Inf. Sci.* **2023**, *66*, 132301. [CrossRef]
15. Yang, Q.; Gu, J.; Wang, D.; Zhang, X.; Tian, Z.; Ouyang, C.; Singh, R.; Han, J.; Zhang, W. Efficient flat metasurface lens for terahertz imaging. *Opt. Express* **2014**, *22*, 25931–35939. [CrossRef]
16. Zou, X.; Zheng, G.; Yuan, Q.; Zang, W.; Chen, R.; Li, T.; Li, L.; Wang, S.; Wang, Z.; Zhu, S. Imaging based on metalenses. *PhotoniX* **2020**, *1*, 2. [CrossRef]
17. Liang, Y.; Koshelev, K.; Zhang, F.; Lin, H.; Lin, S.; Wu, J.; Jia, B.; Kivshar, Y. Bound States in the Continuum in Anisotropic Plasmonic Metasurfaces. *Nano Lett.* **2020**, *20*, 6351–6356. [CrossRef]
18. Jung, H.; Hale, L.L.; Gennaro, S.D.; Briscoe, J.; Iyer, P.P.; Doiron, C.F.; Harris, C.T.; Luk, T.S.; Addamane, S.J.; Reno, J.L.; et al. Terahertz Pulse Generation with Binary Phase Control in Nonlinear InAs Metasurface. *Nano Lett.* **2022**, *22*, 9077–9083. [CrossRef]
19. Beruete, M.; Jáuregui-López, I. Terahertz Sensing Based on Metasurfaces. *Adv. Opt. Mater.* **2020**, *8*, 1900721. [CrossRef]
20. Herbko, M.; Lopato, P. Application of a Single Cell Electric-SRR Metamaterial for Strain Evaluation. *Materials* **2022**, *15*, 291. [CrossRef]
21. Yin, W.; Shen, Z.; Li, S.; Zhang, L.; Chen, X. A Three-Dimensional Dual-Band Terahertz Perfect Absorber as a Highly Sensitive. *Sens. Front. Phys.* **2021**, *9*, 665280. [CrossRef]
22. Zhang, X.; Jin, J.; Wang, Y.; Pu, M.; Li, X.; Zhao, Z.; Gao, P.; Wang, C.; Luo, X. Metasurface-based broadband hologram with high tolerance to fabrication errors. *Sci. Rep.* **2016**, *6*, 19856. [CrossRef] [PubMed]
23. Samanta, S.K.; Pradhan, R.; Syam, D. Theoretical approach to verify the resonance frequency of a square split ring resonator. *J. Opt. Soc. Am. B* **2021**, *38*, 2887. [CrossRef]
24. McNamara, B.P. Measuring the Terahertz Refractive Indes of Boron-Doped Silicon Using a Photoconducting Antenna Terahertz Generator. Master's Thesis, Rochester Institute of Technology, New York, NY, USA, August 2015.
25. Fujii, T.; Ando, A.; Sakabe, Y. Characterization of dielectric properties of oxide materials in frequency range from GHz to THz. *J. Eur. Ceram. Soc.* **2006**, *26*, 1857–1860. [CrossRef]
26. Sun, W.-F.; Wang, X.-K.; Zhang, Y. Measurement of Refractive Index for High Reflectance Materials with Terahertz Time Domain Reflection Spectroscopy. *Chin. Phys. Lett.* **2009**, *26*, 114210. [CrossRef]
27. Li, Z.; Cakmakyapan, S.; Butun, B.; Daskalaki, C.; Tzortzakis, S.; Yang, X.; Ozbay, E. Fano resonances in THz metamaterials composed of continuous metallic wires and split ring resonators. *Opt. Express* **2014**, *22*, 26572–26584. [CrossRef]
28. Herbko, M.; Lopato, P. Double patch sensor for identification of stress level and direction. *Int. J. RF Microw. Comput. Aided Eng.* **2019**, *29*, e21977. [CrossRef]

29. Frei, W. Modeling of Materials in Wave Electromagnetics Problems, Comsol Blog. 2015. Available online: https://www.comsol.com/blogs/modeling-of-materials-in-wave-electromagnetics-problems/ (accessed on 12 January 2023).
30. Vallecchi, A.; Shamonina, E.; Stevens, C.J. Analytical model of the fundamental mode of 3D square split ring resonators. *J. Appl. Phys.* **2019**, *125*, 014901. [CrossRef]

electronics

MDPI

Article

Small Signal Anti-Jamming Scheme Based on a DMA Linear Array under Strong Jamming

Yuankai Wang , Liang Jin *, Yangming Lou and Yinuo Hao

Wireless Communication Technology Office, Information Engineering University, Zhengzhou 450002, China
* Correspondence: liangjin@263.net

Abstract: Considering the difficulty of receiving small signals under strong electromagnetic jamming, this paper proposes a small-signal anti-jamming scheme based on a single dynamic metamaterial antenna (DMA). Our scheme uses the dynamic-adjustable characteristics of the DMA to perform spatial filtering at the antenna radio frequency (RF) front-end, to suppress strong jamming signals in advance and to improve the receiver's ability to receive and demodulate small signals. Specifically, we take the maximization of signal-to-interference-plus-noise ratio (SINR) as the optimization goal, transform the fractional non-convex objective function model into a quasi-convex semi-definite relaxation (SDR) problem, and use the Charnes-Cooper (CC) transform algorithm to find the optimal DMA array-element codeword-state matrix. Simulation results show that DMA has better spatial-beamforming capability than traditional antenna arrays, and the proposed scheme can better resist strong jamming. DMA realizes the effect of digital beamforming at the back end of the traditional communication system, has the advantages of traditional digital-spatial filtering, and further improves the receiver's ability to receive and demodulate small signals.

Keywords: dynamic metamaterial antenna; small signal anti-jamming; matched filtering; signal-to-interference-plus-noise ratio maximization

Citation: Wang, Y.; Jin, L.; Lou, Y.; Hao, Y. Small Signal Anti-Jamming Scheme Based on a DMA Linear Array under Strong Jamming. *Electronics* **2023**, *12*, 1389. https://doi.org/10.3390/electronics12061389

Academic Editors: Naser Ojaroudi Parchin, Mohammad Ojaroudi and Raed A. Abd-Alhameed

Received: 21 February 2023
Revised: 8 March 2023
Accepted: 12 March 2023
Published: 14 March 2023

1. Introduction

With the increasingly complex electromagnetic environment on the battlefield, the dynamic range of the signal has deteriorated sharply under battlefield conditions of large broadband and strong jamming. The jamming patterns faced by radar detection are becoming more and more diverse, and it is becoming more and more difficult to receive target signals. In particular, when a strong jamming signal enters the receiver, it will force the analog-to-digital converter (ADC) to work in the nonlinear region, causing a large number of nonlinear spurs in the received signal, resulting in nonlinear distortion and a small signal. It is submerged in the distortion of the large signal, so that the small signal cannot be received and processed normally. In order to correctly identify small signals, it is often necessary to increase the receiving gain. It can be seen that the wireless signal receiver is a key device in the communication link. Its dynamic performance determines the signal range that the receiver can handle. The lower limit of the dynamic range is the sensitivity of the receiver, that is, the smallest signal that the receiver can normally receive. The upper limit of the dynamic range is determined by the maximum acceptable signal power. When the input signal power is too large, distortion occurs due to the nonlinearity of the receiver. Therefore, the dynamic range of the receiver signal becomes an important bottleneck, restricting the receiving performance of the wireless communication system. A typical wideband receiver includes a low-noise amplifier (LNA), mixer, ADC, and filter; a large dynamic receiver requires the above devices to have good linearity and to reduce the generation of nonlinear distortion [1]. Among them, the ADC is required to have a wide spurious-free dynamic range (SFDR). In order to prevent the working range of the ADC from the linear segment into the nonlinear segment, resulting in nonlinear distortion [2], the power back-off method is used in engineering to deal with this problem. It means to adjust the power of the signal by selecting a lower-gain LNA or by adding an attenuator to ensure that the ADC works in a linear state. However, this operation increases the decision threshold for small signals and reduces the dynamic performance of the receiver. However, the traditional self-adaptive zeroing and anti-jamming technology has been unable to achieve signal alignment due to the dynamic deterioration, or even blockage, of the front end [3]. In fact, high-performance ADC devices have not fundamentally reduced the pressure of back-end digital processing. Therefore, under the premise of not reducing the input-signal power, it will have practical engineering significance to ensure the detection and reception of small signals by suppressing the generation of nonlinear distortion.

At present, research on optimizing wireless signal receivers and improving SFDR signal reception is mainly divided into circuit optimization [4] and digital compensation [5]. In [4], in order to reduce the nonlinear error introduced by capacitance mismatch, the residual voltages of the two sampling stages, before and after, were complementary by exchanging the sampling capacitor. However, the complexity of analog-circuit design is high, and it is difficult for the same circuit to meet the needs of different application scenarios. Therefore, digital-compensation technologies having higher flexibility and better performance have gradually become the focus of research. In existing research, the digital-compensation technology is mainly based on the signal-processing technology of the array antenna [6]. Compared with the traditional antenna, the array antenna has the characteristics of high spatial resolution, long detection distance, and anti-jamming. It is widely used in radar, sonar, wireless communication, space telemetry, and other fields [7–14]. For the electromagnetic jamming problem, the classic digital compensation technology is mainly signal separation. One approach is adaptive-beamforming technology, which can automatically adjust the beam pointing and nulling, according to the direction of the target and the jamming with a specific adaptive criterion, so that the signal output of the array antenna forms a main lobe beam in the direction of the target, and forms a null as low as possible in the jamming direction to suppress the influence of the jamming signal. At the same time, the side-lobe level is minimized to reduce the influence of clutter. Existing studies have applied this technology to enhance target signals, suppress jamming signals, and improve the resolution, energy, and channel utilization of airspace detection. The other was based on the jamming-blocking algorithm and its extended application [15–18]. Its core idea is to use the angle information to construct a jamming-blocking matrix to eliminate the jamming part. In addition, there was an extended-noise subspace algorithm [19], which constructs an extended subspace of strong jamming and noise, and performs conventional direction-of-arrival (DOA) estimation on this basis. In [20], the authors proposed an improved jamming-blocking algorithm, and reconstructed the Toeplitz matrix so that the coherent signal is decoherent, and the jamming signal was suppressed by using the jamming blocking matrix method, but the robustness of the algorithm was not strong.

However, the above-mentioned research methods have been widely used in the reception scenarios of homogeneous-array antennas with electromagnetic jamming, such as anti-jamming. Among them, the homogeneous array-antenna adopts the method of enlarging the array aperture to improve the resolution ability of airspace signals, but when the aperture is limited, it is difficult to achieve the desired signal-separation effect simply by increasing the number of antennas. Currently, both 5G massive multiple-input multiple-output (MIMO) and 6G ultra-massive MIMO face the problem of limited aperture. In fact, the homogeneous array antenna only utilizes the processing gain of the array and ignores the processing gain of the array elements, and the spatial-signal processing-capability of the array antenna is limited [21–23]. At present, an emerging receiver architecture, based on DMA, combines adjustable analog in the hardware level with an array of microstrips, each embedded with configurable radiating-metamaterial elements, the physical properties of which, such as permittivity and magnetic permeability, are dynamically adjustable. By configuring stacks of metamaterials on the surface, they can be tuned to achieve different transformations for transmission, reception, or strike waves. So far, the two main applications of metamaterials have been considered for wireless communications. The first used passive metamaterials as reflectors, also known as reconfigurable intelligent surface (RIS), which generated flexible reflection patterns to change the wireless propagation environment in desired ways [24–26]. Currently, research has been done on holographic imaging [27], intelligent perception [28], wireless communication [29–31], and direction of arrival estimation [32]. The applications are still expanding. Additionally, another emerging application of metamaterials in wireless communications is to exploit their controllable radiation- and reception-patterns as antennas, i.e., the DMA mentioned in this paper. The difference from RIS is that DMA is directly connected to the RF channel, which can change the receiving-response of electromagnetic waves from different directions in real time. It can achieve programmable control of receive-beam patterns through advanced analog signal-processing capabilities. This kind of antenna structure generally consumes less power and costs than traditional array-based architectures. So far, DMA-related research has mainly focused on reducing path loss [33], improving multi-user communication throughput [34–36], and directional beam focusing [37,38], which motivates the study of the effect of DMA on the reception-performance of desired small signals under conditions of strong jamming.

This paper focuses on the matching-reception problem of small signals under the battlefield scenario of strong electromagnetic jamming. Different from the existing antenna-array receiving system, this solution relies on a field-programmable gate array (FPGA) to adjust the DMA array-element codeword-state matrix in real time, which will couple guided waves from in-plane sources into a free space, manipulate the extracted free-space waves on demand, and enable beamforming and steering functions. Each metamaterial element is independently controlled by a positive–intrinsic–negative (PIN) diode to switch the element between the coupled (coded as "1") and uncoupled (coded as "0") states. It can realize the periodic harmonic amplitude and phase-independent adjustment of each metamaterial array element, and can realize the independent adjustment of electromagnetic-wave specific-harmonic pattern and intensity. It is useful to change the electromagnetic-wave response, improve the receiving gain in the direction of the desired small signal, perform spatial-filtering processing in advance at the RF front-end, suppress strong jamming signals, improve the equivalent SFDR of the receiver, and reduce the loss of small-signal information. Considering these advantages of DMA, we innovatively propose a small-signal anti-jamming scheme based on a single DMA, which can perform spatial-filtering and matching reception on multipath channels of multi-stream signals, and improve the receiving gain of desired small signals.

The contributions of the paper are summarized as follows:

First, we established a basic model of small-signal anti-jamming based on single DMA, and the optimal-array element codeword-state matrix of DMA was designed according to the prior direction of arrival and multipath-channel value of the expected small signal and strong jamming, so as to achieve the effect of zero trapping aligned with incoming waves of strong jamming. That is, strong jamming signals will be attenuated, but desired small signals will be boosted at the receiver.

Secondly, in the process of designing the optimal array-element codeword state matrix, our optimization goal is to ensure the maximum SINR of the receiver, so as to maximize the suppression of the impact of strong jamming on the desired small signal. Aiming at the challenging problems caused by non-convex constraints and fractional forms of high-dimensional matrices, we propose a CC transform algorithm, which transforms the fractional-order non-convex objective-optimization model into a quasi-convex SDR problem, to find the optimal DMA array-element codeword-state matrix that maximizes the received SINR to maximize the receive gain of small signals.

Simulation results show that the scheme can achieve higher SINR performance in the face of strong electromagnetic jamming. In addition, the scheme can still improve the effect of small-signal information-loss to a certain extent in the face of strong main lobe jamming.

The rest of the paper is organized as follows: Section 2 introduces the small-signal matching-receiver system model and framework based on a single DMA. Section 3 introduces the small-signal anti-jamming matching-receiving scheme. Section 4 offers the numerical results analysis and performance evaluation. Section 5 draws conclusions.

2. Multi-Stream Signal Matched Reception Scenario

In Figure 1, we consider the small-signal matching receiving system scenario under strong jamming conditions based on DMA, where the antenna RF front-end of the receiver Alice is a DMA with metamaterial array elements, and Bob is a small-signal source. Jammer is a strong jamming signal source. In the process of electromagnetic wave propagation, affected by the terrain, the expected signal transmitted by Bob will reach the DMA of Alice through multiple paths. Each element of the DMA array combines the observed signals at the output port of the microstrip, and feeds the RF chain and ADC through Nyquist rate sampling.

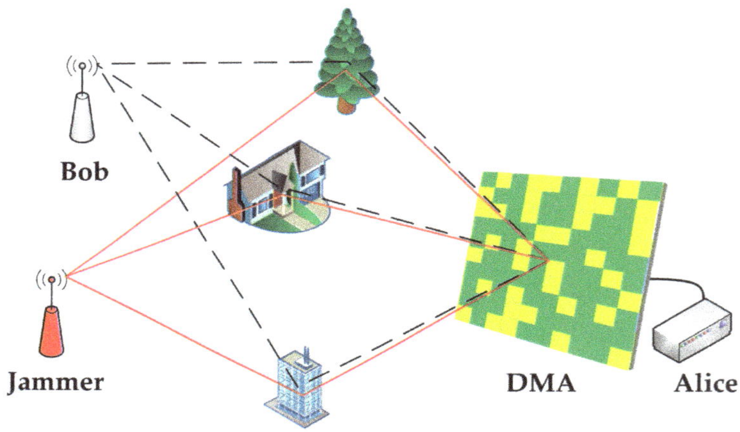

Figure 1. Multi-stream signal matched reception scenario based on DMA.

In this paper, we assume that the directions of arrival of Bob and Jammer are known, and the corresponding multipath channel values have been obtained from the channel estimation results. Then in a single symbol period, Bob's multipath channel estimate can be expressed as:

$$h(\theta_B) \triangleq [h(\theta_{B,1}), \ldots, h(\theta_{B,L_1})]^T, \tag{1}$$

where $h(\theta_B) \in \mathbb{C}^{L_1 \times 1}$ is the multipath channel estimate between Bob and Alice, the multipath number is L_1 and $\theta_B \triangleq [\theta_{B,1}, \ldots, \theta_{B,L_1}]^T$ represents Bob's multipath angle of arrival (AOA) at Alice.

Meanwhile, Jammer's multipath channel estimate can be expressed as:

$$g(\theta_J) \triangleq [g(\theta_{J,1}), \ldots, g(\theta_{J,L_2})]^T, \tag{2}$$

where $g(\theta_J) \in \mathbb{C}^{L_2 \times 1}$ is the multipath channel estimate between Jammer and Alice, the multipath number is L_2, and $\theta_J \triangleq [\theta_{J,1}, \ldots, \theta_{J,L_1}]^T$ represents Jammer's multipath AOA at Alice.

Then the wireless channels between Bob, Jammer, and Alice, respectively, can be expressed as:

$$h_B = \phi^T(\theta_B) h(\theta_B), \tag{3}$$

$$g_J = \phi^T(\theta_J) g(\theta_J), \tag{4}$$

where $\phi(\theta_B) \in \mathbb{C}^{L_1 \times 1}$ is the patten response of DMA at θ_B, $\phi(\theta_J) \in \mathbb{C}^{L_2 \times 1}$ is the patten response of DMA at θ_J and h_B is the equivalent channel between Bob and Alice, g_J is the equivalent channel between Jammer and Alice.

3. Small-Signal Anti-Jamming Model Based on Single DMA

DMA can adjust the response of each metamaterial to incident electromagnetic waves in real time to change the mode, as shown in Figure 2. We consider DMA to be a ULA composed of $N \times 1$ metamaterial elements.

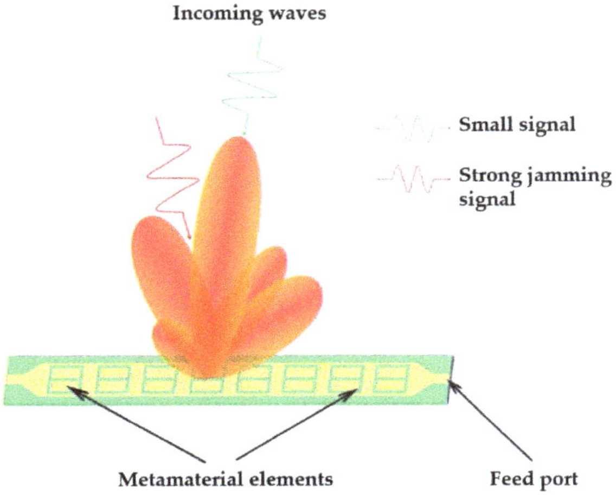

Figure 2. Small-signal anti-jamming model based on single DMA.

The metamaterial array-element pattern of DMA can be written as

$$\phi^T(\theta) = [1 \quad \cdots \quad 1][\Psi(\omega) \odot A(\theta)], \tag{5}$$

$$\begin{aligned} \Psi(\omega) &= \begin{bmatrix} \psi(\omega_1,\theta_1) & \cdots & \psi(\omega_1,\theta_L) \\ \vdots & \ddots & \vdots \\ \psi(\omega_N,\theta_1) & \cdots & \psi(\omega_N,\theta_L) \end{bmatrix} \\ &= \begin{bmatrix} e^{j\varphi(\omega_1,\theta_1)} & \cdots & e^{j\varphi(\omega_1,\theta_L)} \\ \vdots & \ddots & \vdots \\ e^{j\varphi(\omega_N,\theta_1)} & \cdots & e^{j\varphi(\omega_N,\theta_L)} \end{bmatrix}, \end{aligned} \tag{6}$$

where θ represent the direction of arrival AOA of the signal, $\odot$ is for Hadamard product. $A(\theta)$ represents the manifold matrix, where $A(\theta) \triangleq [a(\theta_1), \ldots, a(\theta_L)]$, and $a(\theta) = [1, \ldots, e^{j2\pi d(N-1)\sin\theta/\lambda}]^T$. $\psi(\omega_n, \theta_l)$ represents the amplitude and phase of the incident electromagnetic wave changed by metamaterial elements. Then (5) can be further derived as follows:

$$\phi^T(\theta) = \Omega^T(\omega) A(\theta), \tag{7}$$

where $\Omega(\omega) \triangleq [e^{j\varphi(\omega_1)}, \ldots, e^{j\varphi(\omega_N)}]^T$, $\varphi(\omega_n)$ represent the DMA codeword-state parameter used in the k-th symbol period. The mode of DMA can be customized by adjusting the value of the codeword. Therefore, during the k-th symbol period, assuming the channel remains constant during estimation, the channels of Bob and Jammer can be re-expressed as

$$\begin{aligned} h_B &= \Omega^T(\omega) A(\theta_B) h(\theta_B) \\ g_J &= \Omega^T(\omega) A(\theta_J) g(\theta_J), \end{aligned} \tag{8}$$

4. Small Signal Anti-Jamming Scheme

To simplify the analysis, we assume that the scheme is in a quasi-static wireless communication, and the channel is a fast-fading channel, so the channel remains unchanged during signal-reception and subsequent digital processing.

4.1. Signal Processing

During the k-th symbol period, we set the number of samples at $N_s = 512, k \in [1, K]$. When the k-th pilot symbol is sent, the signal received by DMA can be expressed as

$$y_k = h_B s_{B,k} + g_J s_{J,k} + n_k, \tag{9}$$

where $s_{B,k} \in \mathbb{C}$ is the pilot signal sent by Bob, $s_{J,k} \in \mathbb{C}$ is the pilot signal sent by Jammer. Also $n_k \triangleq [n_{k,1}, \ldots, n_{k,N}]^T$ is the noise received by the N metamaterial elements of the DMA, following the independent and identical distributed (i.i.d.) $\mathcal{CN}(0, \sigma_n^2)$. Arranging the signals received by N metamaterial elements into a column vector, Formula (10) can be re-expressed as

$$y_k = \Omega^T(\omega)(A(\theta_B)h(\theta_B)s_{B,k} + A(\theta_J)g(\theta_J)s_{J,k}) + n_k, \tag{10}$$

where $Y \triangleq [y_1, \ldots, y_K]$, $S_B \triangleq [s_B, \ldots, s_B]$ and $S_J \triangleq [s_{J,1}, \ldots, s_{J,K}]$ represents the pilot sequence respectively. Therefore, in the k-th symbol period, the complex baseband received signal power output by the DMA array can be expressed as

$$P_{out} = \mathbb{E}[\left|y_k {y_k}^H\right|], \tag{11}$$

According to (10), the complex baseband received signal power output by the DMA array can be re-expressed as

$$P_{out} \triangleq \Omega\hat{A}(\theta_B)\hat{P}_B\hat{A}^H(\theta_B)\Omega^H + \Omega\hat{A}(\theta_J)\hat{P}_J\hat{A}^H(\theta_J)\Omega^H + \sigma^2, \tag{12}$$

where $\sigma^2 \in \mathbb{C}$ is the noise power received by DMA. To simplify the analysis, we normalize the signal powers sent by Bob and Jammer to be 1. Then the receiving *SINR* of Alice at the receiving terminal can be written as

$$SINR = \frac{\Omega\hat{A}(\theta_B)\hat{P}_B\hat{A}^H(\theta_B)\Omega^H}{\Omega\hat{A}(\theta_J)\hat{P}_J\hat{A}^H(\theta_J)\Omega^H + \sigma^2.} \tag{13}$$

4.2. Problem Formulation

In order to ensure the reliability of small-signal transmission, we take the maximization of the *SINR* of the received signal as the optimization goal. Therefore, the problem is transformed into optimizing the codeword corresponding to each metamaterial array element of DMA, that is, optimizing $\Omega(\omega)$ to maximize the *SINR*. Therefore, the optimization objective function can be written as

$$P1: \begin{array}{l} \max\limits_{\Omega} \frac{\Omega\Phi\Omega^H}{\Omega\mathbf{R}\Omega^H + \sigma^2} \\ s.t.\ |\Omega_n|^2 = 1,\ n = 1, \ldots, N \end{array}, \tag{14}$$

where ($P1$) is actually a problem of maximizing a fractional non-convex objective function [39]. For such non-convex problems, we can transform it into another more feasible form.

That is $\Omega\Phi\Omega^H = Tr(\Phi\Omega^H\Omega)$ and $\Omega\mathbf{R}\Omega^H(\omega) = Tr(\mathbf{R}\Omega^H\Omega)$, where we define $V = \Omega^H\Omega$ in which $V \geq 0$ and $rank(V) = 1$, $(P1)$ is equivalent to

$$\begin{aligned} P2: \quad & \max_{V} \tfrac{Tr(\Phi V)}{Tr(\mathbf{R}V)+\sigma^2} \\ & s.t.\ V_{n,n} = 1,\ n = 1, \ldots, N \\ & \quad V \geq 0 \\ & \quad rank(V) = 1 \end{aligned} . \tag{15}$$

We will discuss the optimal solution to $(P2)$ in the next section.

4.3. Optimization Algorithm for P2

For this kind of problem, this paper uses SDR to solve it. The term $(P2)$ can be used for the problem of maximizing a fractional non-convex objective function. This usually requires solving using a convex optimization tool. Firstly, we need to remove the $rank(V) = 1$ constraint, so $(P2)$ is simplified as the following SDR models as

$$\begin{aligned} P3: \quad & \max_{V} \tfrac{Tr(\Phi V)}{Tr(\mathbf{R}V)+\sigma^2} \\ & s.t.\ V_{n,n} = 1,\ n = 1, \ldots, N \\ & \quad V \geq 0 \end{aligned} \cdot \tag{16}$$

However, since the objective function of $(P3)$ has a linear fractional structure, an SDR solution cannot be obtained. A classic solution is a binary search algorithm with high computational complexity, which searches for the global optimum by solving a series of SDR solutions. We use the CC transform algorithm to effectively solve this kind of fractional objective function problem, and get the optimal solution of $(P3)$. By introducing a variable $\mu \geq 0$, we transform V as

$$\overline{V} = \mu V. \tag{17}$$

Thus $(P3)$ can be re-expressed as

$$\begin{aligned} P4: \quad & \max_{\overline{V}} Tr(\Phi\overline{V}) \tfrac{Tr(\Phi V)}{Tr(\mathbf{R}V)+\sigma^2} \\ & s.t.\ Tr(\mathbf{R}\overline{V}) + \mu\sigma^2 = \zeta \\ & \quad \overline{V}_{n,n} = \mu,\ n = 1, \ldots, N \\ & \quad \overline{V} \geq 0, \quad \mu \geq 0 \end{aligned} , \tag{18}$$

where ζ is a constant that $\zeta \neq 0$ according to [25]. To solve $(P4)$. We use the CVX toolbox to solve the problem.

5. Numerical Results

In order to establish its input-output relationship, the DMA used in this paper is a single-antenna uniform linear array composed of eight metamaterial array elements. In this paper, we assume that the directions of arrival of Bob and Jammer are known, and the corresponding multipath-channel values have been obtained from the channel estimation results.

5.1. Comparison of Strong Jamming Suppression Effects

In order to verify the performance of the scheme against a strong jamming signal, we first needed to simulate the beam pattern of a single DMA. We set the number of DMA's metamaterial array element bits to $bits = 2$. The metamaterial array-element spacing of the DMA was set to 0.25, the number of snapshots is $N_s = 512$, and the number of metamaterial array elements was $N = 8$. Bob uses a conventional omnidirectional antenna. The simulation scenario architecture is shown in Figure 1. Without loss of generality, the scheme assumes that the channel remains unchanged within a frame period, and each path follows $\mathcal{CN}(0,1)$. The signal-to-noise ratio is $SNR = 10$ dB, and the jamming signal-to-noise ratio is $JNR = 30$ dB. The multipath numbers of the small signal and the strong jamming signal were $L = 3$. The multipath angle distribution was $\theta = [-\frac{\pi}{2}, \frac{\pi}{2}]$. Finally, the single DMA and the homogeneous array antenna have the anti-jamming reception effect on small signals, and the solved pattern is shown in Figure 3.

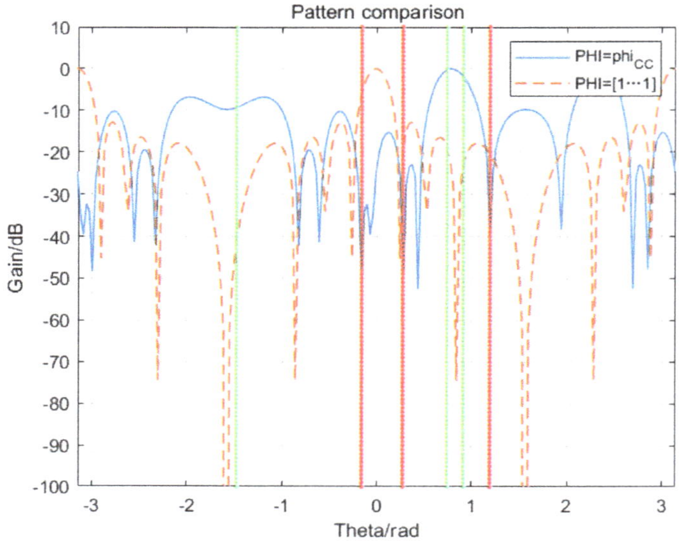

Figure 3. Pattern comparison of the CC algorithm based on DMA and the same-dimensional homogeneous array antenna.

In Figure 3, the green vertical line represents the incoming wave direction of the expected small signal. The red vertical line represents the strong jamming signal. The blue solid line represents the DMA reception pattern of the CC algorithm of this scheme. The orange dotted line represents the receiving pattern of the same-dimension homogeneous array antenna ($PHI = [1 \cdots 1]$). It can be seen that the scheme forms a null in the incoming wave direction of strong jamming, and effectively suppresses strong jamming. Compared with the homogeneous array antenna, the array pattern of DMA is aligned with the direction of arrival of the expected small signal, and the small signal is effectively received, which proves the effectiveness of the scheme.

5.2. Comparison of Small Signal Anti-Jamming Performance

In order to further verify the feasibility of the proposed algorithm, the simulation parameters are basically the same as in Section 5.1, only the value ranges of SNR and JNR are changed respectively. We compared the small-signal anti-jamming performance between the traversal search algorithm ($PHI = phi_{Tra}$) and the traditional adaptive beamforming MVDR algorithm ($PHI = phi_{MVDR}$), and the CC algorithm of this scheme, as shown in Figure 4. It can be seen that the overall range of the SINR of the received signal of the CC algorithm of the scheme is better than that of the other two algorithms, and the SINR of the homogeneous array antenna is the worst. In particular, when the JNR of the strong jamming signal gradually increases, the algorithm can still effectively receive the small signal, and the anti-jamming performance of the small signal is strong, as shown in Figure 4a. This further verifies the effectiveness of this scheme for the strong jamming suppression effect.

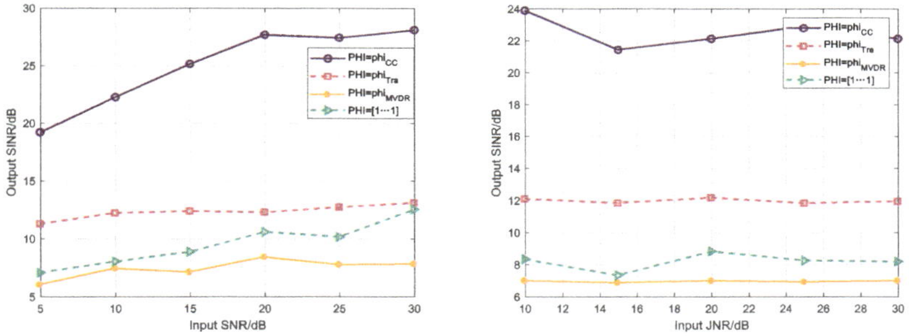

(**a**) SNR vs. SINR (**b**) JNR vs. SINR

Figure 4. The anti-jamming performance of the CC algorithm compared with the traversal search algorithm, MVDR, and homogeneous array antennas. (**a**) SNR. (**b**) JNR.

5.3. Single DMA Performance Analysis

In order to further verify the influence of DMA on the small signal anti-jamming performance, we changed the multipath value in the simulation parameters, the number of bits of the metamaterial element, and the metamaterial-element spacing from the perspective of the array antenna itself. The simulation results are shown in the following chapter.

5.3.1. Effect of the Number of Paths

Under the same simulation conditions, we set the multipath number of small signal and strong jamming signal to $L \in [3, 5, 7]$. The SINR of the received signal after processing is shown in Figure 5. It can be seen that the overall SINR of the signal received by this algorithm is better than that of the traversal search algorithm, and the small-signal receiving effect of the homogeneous array antenna is the worst. From only the three curves of this algorithm, it can be seen that the anti-jamming performance of the number $L = 3$ is the best, followed by $L = 5$, and finally $L = 7$. This shows that the sparsity of the channel in the space will affect the receiving effect of the array antenna. With the increase of the multi-channel number L, the difficulty of aligning the beam of the array antenna to the small signal increases, and only a few of the channels can be matched, thereby reducing the reception performance of the signal DMA.

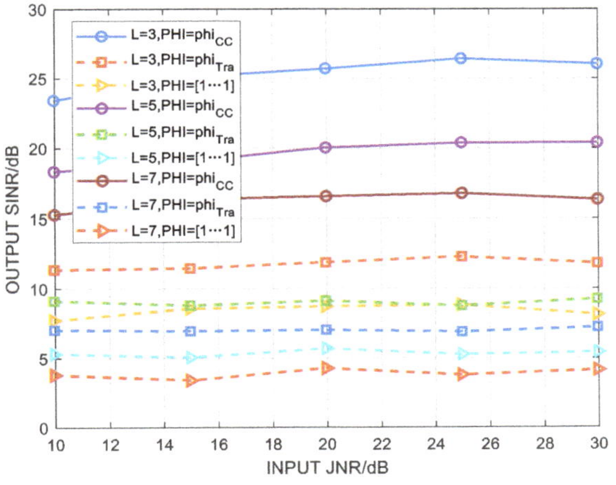

Figure 5. Comparison of the number of paths to the anti-jamming reception performance of different algorithms.

5.3.2. Effect of the Number of Bits in a Single Metamaterial Element

In order to illustrate the effect of the algorithm in this paper on improving small-signal information loss, it is compared with the traversal search algorithm, the traditional adaptive beamforming MVDR algorithm, and the homogeneous-array antenna. The number of bits of each metamaterial element is adjusted to $bits \in [1,2,3]$, and the other parameters are the same as in Section 5.1. The SINR of the received signal after processing is shown in Figure 6. It can be seen that by increasing the number of bits of a single metamaterial element, this solution can slightly improve the receiving SINR of small signals, but the trend is not obvious. This shows that although the increase in the number of bits can enable metamaterial elements to have more receiving states, the signal-processing capability of a single DMA line array is limited, while the shape and parameters of the DMA should be configured according to specific scenarios. Even so, this solution still has good anti-jamming performance against strong jamming, and the method of increasing the signal-processing gain by increasing the number of bits cannot be denied, just like the suboptimal traversal search algorithm in Figure 6. However, the homogeneous array antenna only utilizes the processing gain of a single array and ignores the processing gain of the array element itself. Therefore, the method of increasing the number of bits does not essentially improve the processing performance of the homogeneous array antenna for airspace signals.

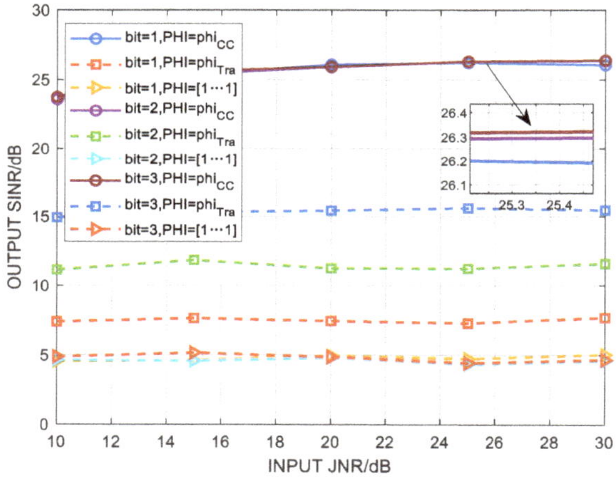

Figure 6. Comparison of the anti-jamming performance of different algorithms based on the number of bits in a single metamaterial element.

5.3.3. Effect of Metamaterial Element Spacing

Under the same simulation conditions, we set the multipath number of small signal and strong jamming signal to $d/\lambda \in [1/2, 1/3, 1/4]$. The SINR of the received signal after processing is shown in Figure 7. It can be seen from Figure 7 that the element spacing can also affect the SINR of the array antenna to the received signal. On the whole, when the metamaterial element spacing is further reduced, the SINR of DMA for small signal reception shows a downward trend, which indicates that in the strong jamming scenario, the metamaterial element spacing should not be too small, and the spacing should be adjusted according to the actual scene to improve each metamaterial element's perception and observation ability. The signal can further improve the receiving-gain of the small signal.

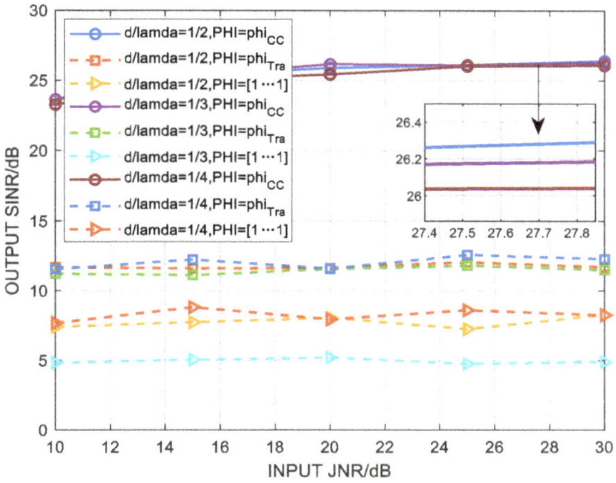

Figure 7. Comparison of anti-jamming performance of different algorithms based on metamaterial element spacing.

5.3.4. Effect of Strong Main Lobe Jamming

In order to further analyze the performance of this scheme against strong main-lobe jamming, we set strong jamming signal sources in the range of 2° deviation of small signal incoming direction. We confirmed that the jamming signal enters the main lobe range, and that the other simulation conditions were the same as Section 5.1. At this time, the SINR of the received signal is shown in Figure 8. It can be seen that when the strong jamming signal enters the main lobe range, the reception-gain of all schemes for small signals is greatly reduced, and the small signal information loss is large. However, CC algorithm can still improve the effect of small-signal information loss to a certain extent, which proves the feasibility of the proposed algorithm.

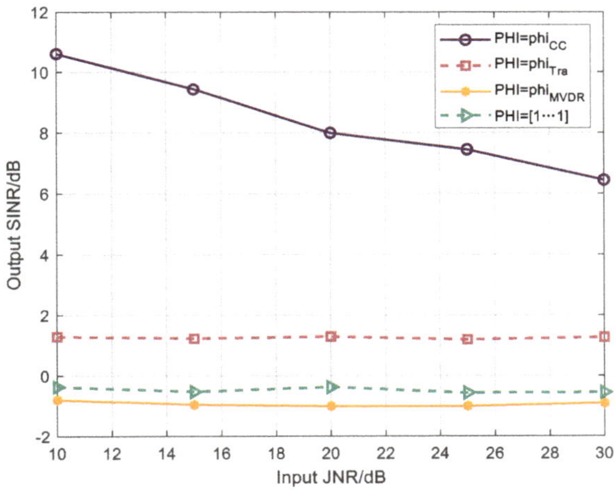

Figure 8. Comparison of small signal anti-jamming performance under strong main lobe jamming.

6. Discussion

In summary, the scheme proposed in this paper has good small-signal anti-jamming performance. Compared with the traversal search algorithm and the traditional adaptive beamforming MVDR algorithm, the CC algorithm we adopt not only improves the receiving gain of small signals under strong jamming, but also can be used even when strong jamming enters the range of the main lobe. To a certain extent, the information loss of small signals is reduced, which proves the feasibility and superiority of this scheme.

In addition, we also conducted an in-depth analysis of the parameter characteristics of single DMA. This solution can effectively improve the receiving SINR of small signals when dealing with strong far-field electromagnetic jamming scenarios, as shown in Figure 5. This method of adjusting the number of DMA bits or the metamaterial element spacing can improve the anti-jamming effect of small signals to a certain extent, and the specific parameters should be reasonably configured according to the actual scene.

7. Conclusions

This paper studies the problem of anti-jamming matching-reception of small-signal anti-jamming based on single DMA under strong electromagnetic jamming. In view of the limited dynamic range of signal processing of traditional antenna receivers, and the fact that the small signal cannot be received normally in the scene of strong electromagnetic jamming on the battlefield, this paper proposes a small-signal anti-jamming reception-enhancement model based on a single DMA to ensure the maximum SINR as the optimization goal, transforming the non-convex optimization model into a quasi-convex SDR problem. We use the CC transform algorithm to find the optimal DMA array-element codeword-state matrix that maximizes the received SINR. The simulation results show that DMA has more advantages than traditional array antennas. With better spatial beamforming capability, the scheme can resist strong jamming. Even if the jamming intensity is high, the scheme can form attenuation in the jamming direction, and form a main lobe in the direction of the desired small signal, and the small signal can obtain a large receiving gain.

Additionally, this paper provides a feasible solution for the direction of small-signal anti-jamming in the field of strong electromagnetic countermeasures. In the future, we will be able to study the anti-jamming performance of multiple DMA line-arrays, or two-dimensional arrays for wireless communication, from both spatial and temporal dimensions in combination with the digital coding and control characteristics of DMA. In addition, DMA is coded and regulated based on FPGA, so it can be matched to and compatible with existing wireless communication software and hardware architectures. Compared with traditional antenna arrays, the communication overhead is further reduced. These advantages of DMA make it possible to innovate and upgrade existing wireless communication systems.

Author Contributions: Y.W. and L.J. proposed the methodology; Y.W. conducted the theoretical analysis as well as the simulation verification; Y.L. and L.J. managed and coordinated responsibility for the research activity planning and execution; Y.W. wrote the original draft, which was reviewed and edited by Y.L. and Y.H. All authors have read and agreed to the published version of the manuscript.

Funding: This work is supported by the National Natural Science Foundation of China (No. U22A2001), the National Key Research and Development Program of China (2022YFB2902202), and the Song Shan Laboratory Program (Included in the management of Major Science and Technology Program of Henan Province): 221100211300-03.

Data Availability Statement: Not applicable.

Conflicts of Interest: The authors declare no conflict of interest.

References

1. Wang, N. Multi-channel Signal Receiving for Improving Spurious-Free Dynamic Range. *Telecommun. Eng.* **2021**, *61*, 603–607.
2. Pedro, C.; Nuno, B.; Kate, A. Evaluation of Nonlinear Distortion in ADCs Using Multisines. In Proceedings of the 2008 IEEE MTT-s International Microwave Symposium Digest, Atlanta, GA, USA, 15–20 June 2008; pp. 1433–1436.
3. Lin, Z.; Chen, Y.; Liu, X.; Jiang, R.; Shen, B. Adaptive Beamforming Design of Planar Arrays Based on Bayesian Compressive Sensing. *IEEE Sens. J.* **2021**, *21*, 5185–5194. [CrossRef]
4. Yun, C.; Gray, P.; Nikolic, B. A 14-b 12-MS/s CMOS Pipeline ADC with over 100-dB SFDR. *IEEE J. Solid-State Circuits* **2004**, *39*, 2139–2151.

5. Zou, Y.; Wu, T.; Peng, Y.; Sun, Y.; Liu, Y.; Zhang, C. Time Interleaved Sampling ADC Mismatch Error Compensation Method Based on Offline Estimation and Online Compensation. *J. Electron. Inf.* **2019**, *41*, 226–232.
6. Irazoqui, R.; Fulton, C. Spatial Interference Nulling Before RF Frontend for Fully Digital Phased Arrays. *IEEE Access* **2019**, *7*, 151261–151272. [CrossRef]
7. Zhang, K.; Guo, C. Research Status and Development of Adaptive Beamforming Technology. In Proceedings of the 11th China Satellite Navigation Annual Conference, Chengdu, China, 20–22 April 2020; pp. 33–36.
8. Su, Y.; Wang, T.; Lin, X. A Novel Decoupled Narrowband Robust Adaptive Beamforming Based on The Alternating Direction Method of Multipliers in A Noisy Channel. *IET Radar Sonar Navig.* **2020**, *14*, 637–642. [CrossRef]
9. Fu, W.; Jiang, D. Radar Wideband Digital Beamforming Based on Time Delay and Phase Compensation. *Int. J. Electron.* **2018**, *105*, 1144–1158. [CrossRef]
10. Jin, Z.; Yu, W.; Dan, S. Torque Identification Based on LMS Adaptive Delay Inverse System Method. *J. Meas. Eng.* **2020**, *8*, 62–71.
11. Lin, Z.; Lin, M.; Champagne, B.; Zhu, W.P.; Al-Dhahir, N. Secrecy-energy efficient hybrid beamforming for satellite-terrestrial integrated networks. *IEEE Trans. Commun.* **2021**, *69*, 6345–6360. [CrossRef]
12. Lin, Z.; Lin, M.; De Cola, T.; Wang, J.B.; Zhu, W.P.; Cheng, J. Supporting IoT with rate-splitting multiple access in satellite and aerial-integrated networks. *IEEE Internet Things J.* **2021**, *8*, 11123–11134. [CrossRef]
13. Lin, Z.; Niu, H.; An, K.; Wang, Y.; Zheng, G.; Chatzinotas, S.; Hu, Y. Refracting RIS-aided hybrid satellite-terrestrial relay networks: Joint beamforming design and optimization. *IEEE Trans. Aerosp. Electron. Syst.* **2022**, *58*, 3717–3724. [CrossRef]
14. Lin, Z.; Lin, M.; Champagne, B.; Zhu, W.P.; Al-Dhahir, N. Secure beamforming for cognitive satellite terrestrial networks with unknown eavesdroppers. *IEEE Syst. J.* **2020**, *15*, 2186–2189. [CrossRef]
15. Yang, Y.; Zhang, Y.; Long, Y. Wideband Sparse Spatial Spectrum Estimation Using Matrix Filter with Nulling in A Strong jamming Environment. *J. Acoust. Soc. Am.* **2018**, *143*, 3891–3898. [CrossRef] [PubMed]
16. Gong, J.; Lou, S.; Guo, Y. DOA Estimation Method of Weak Sources for An Array Antenna under Strong jamming Conditions. *Int. J. Electron.* **2018**, *105*, 530–534. [CrossRef]
17. Chen, H.; Su, H. A New Approach to Estimate DOA in Presence of Strong Jamming/Signal Suppression. *Acta Electron. Sin.* **2006**, *34*, 530–534.
18. Fang, Q.; Jin, M.; Liu, W.; Han, Y. DOA Estimation for Sources with Large Power Differences. *Int. J. Antennas Propag.* **2021**, *2021*, 8862789. [CrossRef]
19. Zhang, J.; Liao, G.; Zhang, J. DOA Estimation Based on Extended Noise Subspace in The Presence of Strong Signals. *Syst. Eng. Electron.* **2009**, *31*, 1276–1283.
20. Li, H.; Guo, Y.; Gong, J.; Tian, B. DOA Estimation Method for Coherent Weak Sources in the Presence of Strong Jamming. *Mod. Radar* **2012**, *34*, 45–49.
21. Jin, L.; Lou, Y.; Xu, X.; Zhong, Z.; Wang, H. Separating Multi-Stream Signals Based on Space-Time Isomerism. In Proceedings of the 2020 International Conference on Wireless Communications and Signal Processing (WCSP), Nanjing, China, 21–23 October 2020; pp. 418–423.
22. Zhang, L.; Chen, X.; Liu, S.; Zhang, Q.; Zhao, J.; Dai, J.; Bai, G. Space-Time-Coding Digital Metasurface. In Proceedings of the 2019 Thirteenth International Congress on Artificial Materials for Novel Wave Phenomena (Metamaterials), Rome, Italy, 16–21 September 2019; pp. 128–130.
23. Zhao, J.; Yang, X.; Dai, J. Programmable Time-domain Digital-Coding Metasurface for Non-linear Harmonic Manipulation and New Wireless Communication Systems. *Natl. Sci. Rev.* **2019**, *6*, 55–62. [CrossRef]
24. Yan, W.; Yuan, X.; He, Z.; Kuai, X. Passive Beamforming and Information Transfer Design for Reconfigurable Intelligent Surfaces Aided Multiuser MIMO Systems. *IEEE J. Sel. Areas Commun.* **2020**, *38*, 1793–1808. [CrossRef]
25. Ma, Z.; Chen, X.; Xiao, M.; Karagiannidis, G.; Fan, P. Interference Control for Railway Wireless Communication Systems: Techniques, Challenges, and Trends. *IEEE Veh. Technol. Mag.* **2020**, *15*, 51–58. [CrossRef]
26. Zhang, L.; Chen, X.Q.; Liu, S.; Zhang, Q.; Zhao, J.; Dai, J.Y.; Bai, G.D.; Wan, X.; Cheng, Q.; Castaldi, G.; et al. Space-time-coding digital metasurfaces. *Nat. Commun.* **2018**, *9*, 4334. [CrossRef] [PubMed]
27. Li, L.; Cui, T.J.; Ji, W.; Liu, S.; Ding, J.; Wan, X.; Li, Y.B.; Jiang, M.; Qiu, C.-W.; Zhang, S. Electromagnetic reprogrammable coding-metasurface holograms. *Nat. Commun.* **2017**, *8*, 197. [CrossRef]
28. Ma, Q.; Bai, G.D.; Jing, H.B.; Yang, C.; Li, L.; Cui, T.J. Smart metasurface with self-adaptively reprogrammable functions. *Light Sci. Appl.* **2019**, *8*, 98. [CrossRef] [PubMed]
29. Cui, T.J.; Liu, S.; Bai, G.D.; Ma, Q. Direct transmission of digital message via programmable coding metasurface. *Research* **2019**, *2019*, 2584509. [CrossRef]
30. Dai, J.Y.; Tang, W.K.; Zhao, J.; Li, X.; Cheng, Q.; Ke, J.C.; Chen, M.Z.; Jin, S.; Cui, T.J. Wireless communications through a simplified architecture based on time-domain digital coding metasurface. *Adv. Mater. Technol.* **2019**, *4*, 1900044. [CrossRef]
31. Zhang, L.; Chen, M.Z.; Tang, W.; Dai, J.Y.; Miao, L.; Zhou, X.Y.; Jin, S.; Cheng, Q.; Cui, T.J. A wireless communication scheme based on space-and frequency-division multiplexing using digital metasurfaces. *Nat. Electron.* **2021**, *4*, 218–227. [CrossRef]
32. Lin, M.; Xu, M.; Wan, X.; Liu, H.; Wu, Z.; Liu, J.; Deng, B.; Guan, D.; Zha, S. Single sensor to estimate DOA with programmable metasurface. *IEEE Internet Things J.* **2021**, *8*, 10187–10197. [CrossRef]

33. Zhang, H.; Shlezinger, N.; Guidi, F.; Dardari, D.; Imani, M.F.; Eldar, Y.C. Beam focusing for multi-user MIMO communications with dynamic metasurface antennas. In Proceedings of the ICASSP 2021–2021 IEEE International Conference on Acoustics, Speech and Signal Processing (ICASSP), Toronto, ON, Canada, 6–11 June 2021; pp. 4780–4784.
34. Shlezinger, N.; Alexandropoulos, G.C.; Imani, M.F.; Smith, D.R. Dynamic Metasurface Antennas for 6G Extreme Massive MIMO Communications. *IEEE Wirel. Commun.* **2021**, *28*, 106–113. [CrossRef]
35. Shlezinger, N.; Dicker, O.; Eldar, Y.C.; Yoo, I.; Imani, M.F.; Smith, D.R. Dynamic Metasurface Antennas for Uplink Massive MIMO Systems. *IEEE Trans. Commun.* **2019**, *67*, 6829–6843. [CrossRef]
36. Yoo, I.; Smith, D.R. Dynamic Metasurface Antennas for Higher-Order MIMO Systems in Indoor Environments. *IEEE Wirel. Commun. Lett.* **2020**, *9*, 1129–1132. [CrossRef]
37. Dardari, D. Communicating with large intelligent surfaces: Fundamental limits and models. *IEEE J. Sel. Areas Commun.* **2020**, *38*, 2526–2537. [CrossRef]
38. Wu, G.B.; Dai, J.Y.; Cheng, Q.; Cui, T.J.; Chan, C.H. Sideband-free space–time-coding metasurface antennas. *Nat. Electron.* **2022**, *5*, 808–819. [CrossRef]
39. Stanley, Z. Programming with Linear Fractional Functionals. *Nav. Res. Logist. Q.* **1968**, *15*, 449–451.

electronics

MDPI

Communication

Negative Group Delay Metamaterials Based on Split-Ring Resonators and Their Application

Zheng Liu, Jian Zhang *, Xue Lei, Jun Gao, Zhijian Xu and Tianpeng Li

National Digital Switching System Engineering and Technological R&D Center, Zhengzhou 450000, China
* Correspondence: zhang_xinda@126.com

Abstract: In this report, negative group delay (NGD) metamaterials based on split-ring resonators (SRRs) are discussed. A theoretical analysis is proposed to calculate the equivalent circuit parameters, NGD values, and S21 amplitudes of two types of SRRs. Metamaterials made from tantalum nitride are simulated, and the parameters of the two types of SRRs are discussed. Prototypes of metamaterials were fabricated and tested. Measured real-world results were found to be consistent with theoretical and simulated predictions. For EC-SRR, a negative group delay of up to −0.1 ns was achieved at 12–13 GHz. For SR-SRR of the same size as the out ring of EC-SRR, a negative group delay of up to −0.04 ns was achieved, with a loss lower than 2.7 dB. The proposed SRRs were applied to continuous transverse stub (CTS) antenna to reduce the beam walk. The simulation shows that the beam walk can be reduced using the proposed metamaterial.

Keywords: negative group delay; split-ring resonator; tantalum nitride

1. Introduction

In recent years, there has been a great deal of work devoted to metamaterials. Metamaterials have many excellent properties: low cost, easy fabrication, excellent properties, and low profile. Among the metamaterials, split-ring resonators (SRRs) are the most commonly used structures [1]. In [2], Chen et al. proposed far-field decoupling of a two-element antenna transceiving system. In [3], Huang et al. designed a multi-band SRR antenna array. In [4–6], a negative group delay (NGD) performance was achieved. In [7], Liang et al. studied strong field enhancement using SRRs. In [8], Lalbakhsh proposed an all-metal wideband metamaterial.

Since Brillouin et al. discovered NGD performance in the 1906s [9], NGD devices have been used in many systems. In [10], NGD devices were used to achieve a flat group delay response when cascaded with conventional microwave components. In [11], NGD devices were used to achieve power dividers. In [12], an amplifier using the NGD circuit to achieve group delay equalization was proposed. NGD in metamaterials was observed in a 1999 report by Pendry et al. [1], and was further discussed by Woodley et al. [5], but performance metrics such as NGD values and the relationship between parameters and loss have not been extensively studied.

In this study, equivalent circuits were proposed for both a single-ring SRR (SR-SRR) and an edge-coupled SRR (EC-SRR), both with negative group delay. The equivalent circuit parameters, especially resistance, were given and the NGD performances of these proposed circuits were then analyzed; prototypes made from Tantalum Nitride were then fabricated and tested. Additionally, the application of the SRRs to CTS antenna is discussed.

This paper is organized as follows: Section 2 describes the proposed equivalent circuit and details our analysis of its SRRs; Section 3 describes the simulated and measurement results of the SRRs; Section 4 describes the application of the SRRs; the main conclusions of this study are then summarized in Section 5.

Citation: Liu, Z.; Zhang, J.; Lei, X.; Gao, J.; Xu, Z.; Li, T. Negative Group Delay Metamaterials Based on Split-Ring Resonators and Their Application. *Electronics* **2023**, *12*, 1064. https://doi.org/10.3390/electronics12041064

Academic Editors: Naser Ojaroudi Parchin, Mohammad Ojaroudi and Raed A. Abd-Alhameed

Received: 7 February 2023
Revised: 15 February 2023
Accepted: 16 February 2023
Published: 20 February 2023

2. Theory

2.1. Stucture and Equivalent Circuit

The structures of two types of SRRs are shown in Figure 1 below.

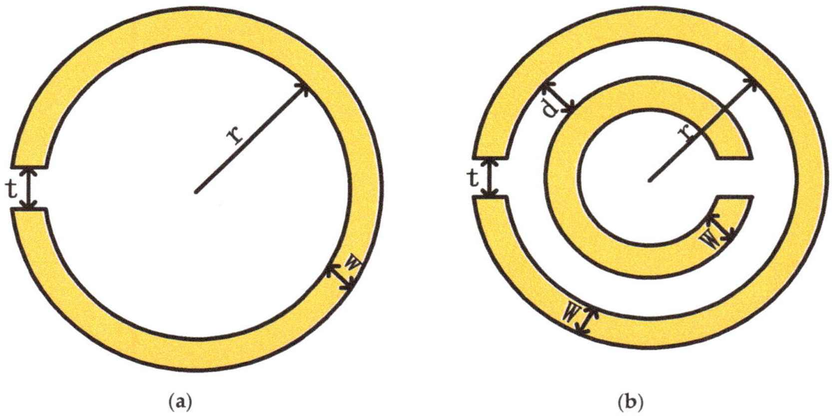

Figure 1. Diagrams of two types of SRRs: (**a**) single-ring SRR (SR-SRR); (**b**) edge-coupled SRR (EC-SRR).

These two kinds of SRRs have been studied by many researchers [1,13], and the same equivalent circuit can be obtained for them, as shown in Figure 2. A transmission line with a series resistor–inductor–capacitor (R–L–C) circuit was used to analyze them, which was consistent with our simulation and experiment setups. The inductance and capacitance (LC) parameters of SR-SRRs can be obtained [13,14] by:

$$L = \frac{\mu \pi r^2}{\mathrm{h}} \tag{1}$$

$$C_{SR} = \frac{\varepsilon \mathrm{Wh}}{\mathrm{t}} \tag{2}$$

where μ is the permeability, ε is the permittivity, r is the inner radius of the split ring, W is the width of the split ring, h is the height of the split-ring resonator, and t is the gap width. The inner radius r can be obtained using the surface resistance multiplied by the area fraction, as shown below:

$$R = R_s P = \sqrt{\frac{\omega\mu}{2\sigma}} \frac{C_{fr}\pi(r + W)^2}{a^2} \tag{3}$$

where a is the period of the metamaterial, σ is the conductivity of the surface, and $R_s = \sqrt{\frac{\omega\mu}{2\sigma}}$ is the surface resistance. P_{fr} is the area fraction constant, which can be set to 2 for the two surfaces of the metamaterial, and can be optimized if losses from other parts of the SRRs are considered.

The inductor, L, and resistor, R, of an EC-SRR are the same as for a SR-SRR, and the capacitor C can be obtained by:

$$C_{EC} = \frac{\varepsilon}{\pi} \ln\left(\frac{2\mathrm{W}}{\mathrm{d}}\right) \tag{4}$$

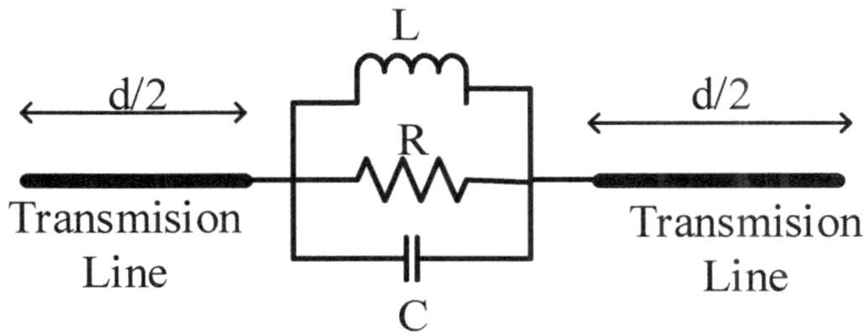

Figure 2. Equivalent circuit for split-ring resonators.

It should be noted that with a sufficiently large d value, an EC-SRR becomes equivalent to a SR-SRR, and that with a small d value, coupled capacitance is the main factor that influences the resonant frequency.

2.2. Negative Group Delay and Loss

S_{11} and S_{21} can be obtained from an ABCD matrix of the equivalent circuit as:

$$S_{11} = \frac{S_R^n + jS_I^n}{S_R^d + jS_I^d} S_{21} = \frac{2}{S_R^d + jS_I^d} \tag{5}$$

where

$$S_I^d = 2\sin(\beta d) + \frac{wLR\left(R - w^2\,\mathrm{LCR}\right)\cos(\beta d)}{\left(R - w^2\mathrm{LCR}\right)^2 + (wL)^2} + \frac{(wL)^2 r\,\sin(\beta d)}{\left(R - w^2\mathrm{LCR}\right)^2 + (wL)^2}$$

$$S_R^d = 2\cos(\beta d) - \frac{wLR\left(R - w^2\,\mathrm{LCR}\right)sin(\beta d)}{\left(R - w^2\mathrm{LCR}\right)^2 + (wL)^2} + \frac{(wL)^2 R\,\cos(\beta d)}{\left(R - w^2\mathrm{LCR}\right)^2 + (wL)^2}$$

$$S_R^n = \frac{wLR\left(R - w^2\,\mathrm{LCR}\right)\cos(\beta d)}{\left(R - w^2\mathrm{LCR}\right)^2 + (wL)^2} + \frac{(wL)^2 R\,\sin(\beta d)}{\left(R - w^2\mathrm{LCR}\right)^2 + (wL)^2}$$

$$S_I^n = \frac{(wL)^2 R\,\cos(\beta d)}{\left(R - w^2\mathrm{LCR}\right)^2 + (wL)^2} - \frac{wLR\left(R - w^2\,\mathrm{LCR}\right)sin(\beta d)}{\left(R - w^2\mathrm{LCR}\right)^2 + (wL)^2}$$

In this formula, d is the total length of the transmission line, β is the propagation constant, R is the resistance, L is the inductance, C is the capacitance, and w is the angular frequency. The amplitudes of S11 and S21 can be simplified at the resonant frequency as:

$$|S_{11}| = \frac{R}{2+R}|S_{21}| = \frac{2}{2+R} \tag{6}$$

The total loss can be obtained by:

$$Loss = 1 - |S_{21}|^2 - |S_{22}|^2 = \frac{4R}{(2+R)^2} \tag{7}$$

The loss increases proportionately with R when R is lower than 2. It should be noted that the loss is included with the return loss, but for transmission metamaterials the transmission power is more important than the absolute loss; therefore, S21 which is the ratio between input power and output power, is the critical parameter to evaluate.

S21 phase and group delays can be simplified at the resonant frequency as:

$$p_{s21} = \arctan(-\frac{S_I}{S_R}) \tag{8}$$

$$GD = \frac{\partial p_{s21}}{\partial \omega}|_{w=w_0} = \frac{\partial \beta}{\partial \omega}|_{w=w_0} - \frac{2CR^2}{2+R} \tag{9}$$

It can be deduced from Equations (6) and (9) that the amplitude of S21 is determined by *R*, and the group delay is determined by *L* and *R*.

3. Simulation and Measurement Results

3.1. Simulation Results

Simulations were conducted using the Ansys HFSS suite (Ansys Inc., Canonsburg, PA, USA). The simulation setup is shown in Figure 3. The TE_{00} mode was used to excite the SRRs; this is same as the transformation metamaterial setup. Teflon substrate was used to as a supporting structure which is widely used and has good electromagnetic properties and the height of which is same as the SRRs. The periodic boundary and the input floquet port are adopted which are used to simulate infinitely large structure and the edge effect can been ignored. The material used for SRRs is tantalum nitride (TaN), and the conductivity is 7400 S/m. The skin depth of the TaN at 15 GHz is 0.04 mm. Therefore, the smallest possible size of the structure should be greater than 0.04 mm and the solving inside threshold of its conductivity was set at 10,000 S/m. It should be noted that the TE_{00} mode was used in simulating the metamaterials with periodic boundaries, which is the normal input field mode of metamaterials; however, in measurement setup, the mode of the rectangular waveguide was the TE_{10} mode, which may cause a slight difference in resonant frequency compared with TE_{00} mode excitation.

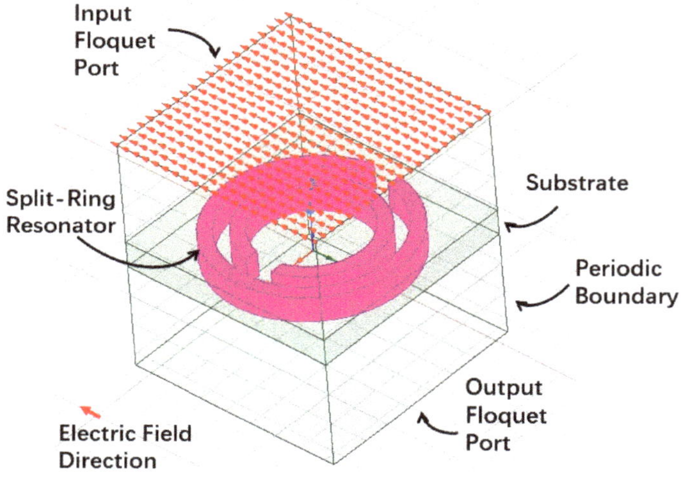

Figure 3. Simulation setup.

3.1.1. SR-SRR

It can be seen from the simulation results (Figure 4) that as w and r increase, the resonant frequency is reduced, which is due to the increase in *L* and *C*. The loss increases with the reduction in resonant frequency, which is due to the fact that when r and w increase, the area of the ring increased as well, which in turn increases the loss. When t increases, the *L* and *C* reduced and then the resonant frequency increases as well. Although the increase in t can reduce the area of the ring to some extent, the increase in the frequency of the surface resistance mainly influences the loss. Additionally, according to Equation (9), the increase in the surface resistance will increase the NGD value.

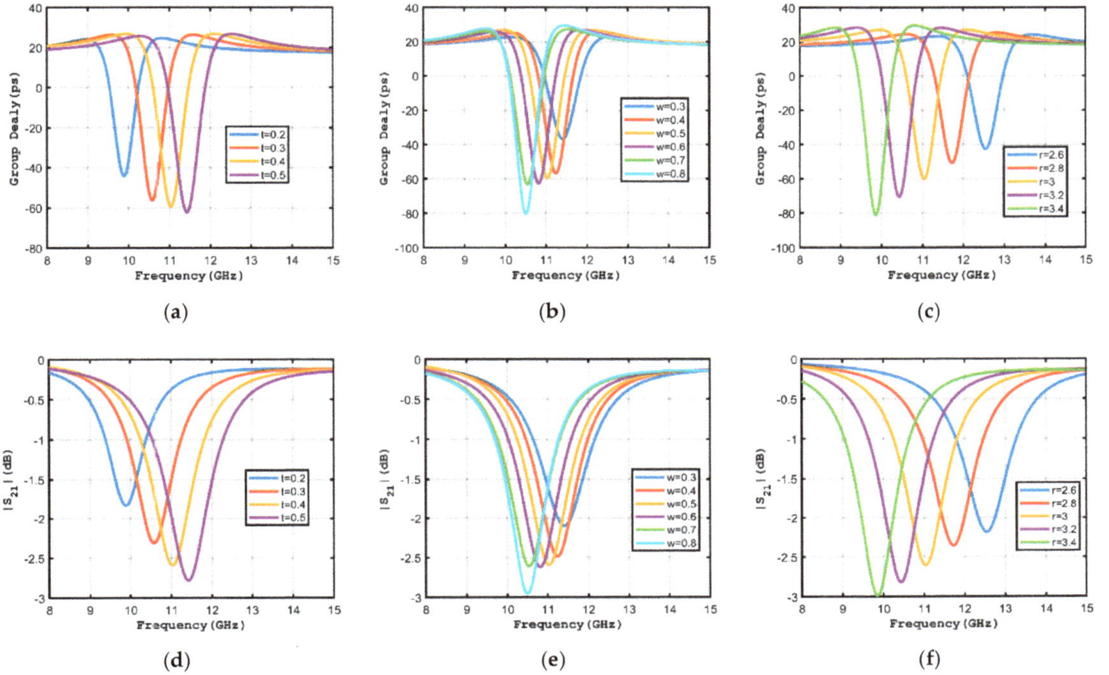

Figure 4. Simulated group delay with different (**a**) t, (**b**) w, (**c**) r, and S21 amplitudes with different amounts of (**d**) t, (**e**) w, (**f**) r.

3.1.2. EC-SRR

The performances of r, t, and w for EC-SRRs (Figure 5) are the same as for SR-SRRs, which can be seen from the equivalent circuit of EC-SRRs where the L term, and part of the C term, are decided by the outer ring. The only difference between the two is in the d parameter. When d increases, the coupled capacitance between the rings is reduced and the resonant frequency is reduced, which in turn decreases the surface resistance. Additionally, according to Equation (9), the decrease in surface resistance will reduce the NGD value.

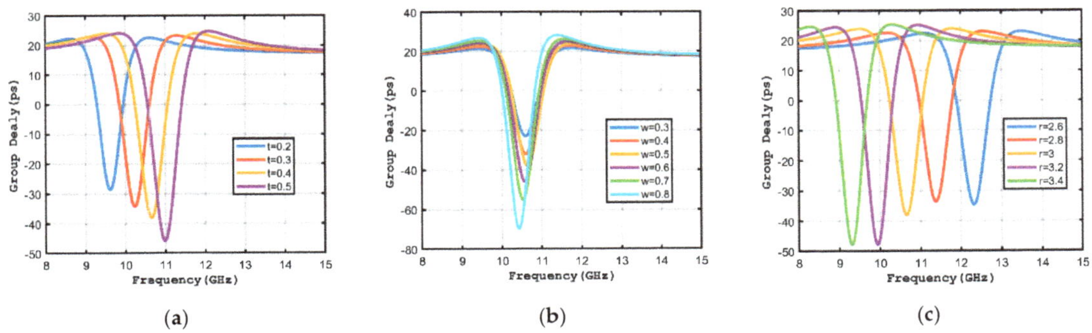

Figure 5. *Cont.*

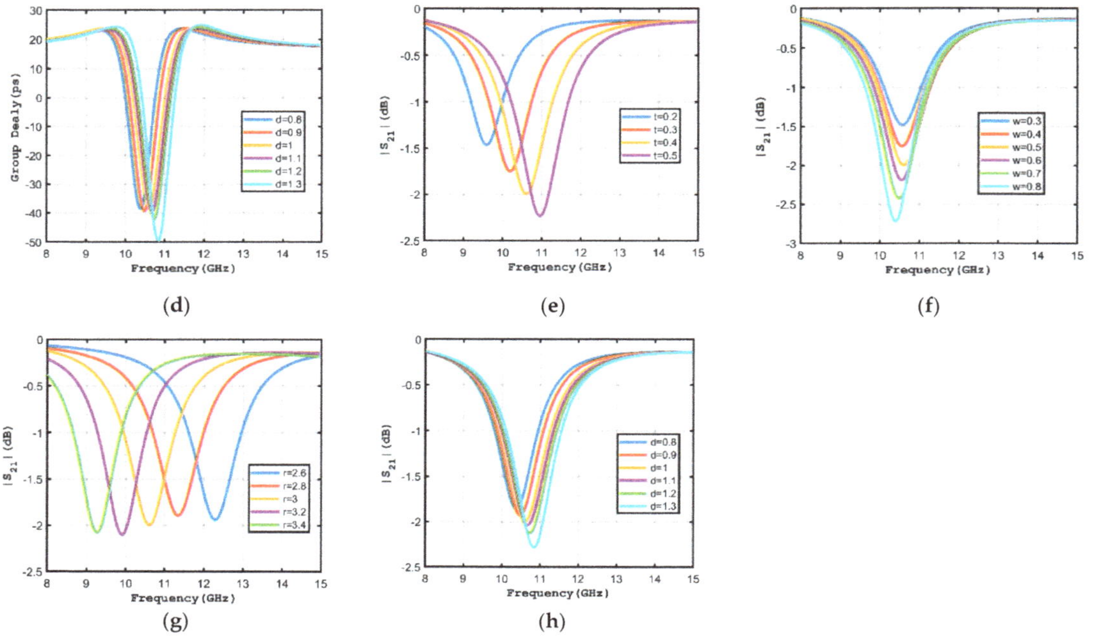

Figure 5. Simulated group delay with different (**a**) t, (**b**) w, (**c**) r, (**d**) r, and S21 amplitudes with different amounts of (**e**) t, (**f**) w, (**g**) r, (**h**) d.

3.2. Measured Results

The prototypes of the proposed structures were fabricated (Figure 6a) and tested using an Agilent E8364C network analyzer (Agilent Technologies Inc., Santa Clara, CA, USA) (Figure 6b). Due to the high hardness and high machining accuracy requirement of the SRRs, laser cutting was used to process the TaN material, which has a low cost in mass production. Additionally, the Teflon substrate was fabricated using computer numerical control (CNC) machining.

Figure 6. (**a**) Fabricated prototype structures and (**b**) measurement setup.

The metamaterials were put into the same type of WR75 rectangular waveguide used [1]. in their 1999 report which is a widely used measurement setup. The parameters used were: t = 0.5 mm, r = 1.3 mm, w = 0.5 mm, h = 0.3 mm, and d = 0.5 mm. The simulated and measured results are shown in Figures 7 and 8, respectively, to illustrate the consistency between them. The simulation was performed in the waveguide condition. The small observed differences may be attributable to the Teflon substrate used in the real-world test, of which the dielectric constant is not stable. The cut-off frequency of WR75 waveguide is 10 GHz, and the S21 and group delay performance below 10 GHz should be ignored. It can be seen from Figures 7 and 8 that with the same outer ring dimension, the S21 of the EC-SRR was lower than that of the SR-SRR, due to the reduction in the resonant frequency. The measured S11 was also consistent with the theoretical and simulated predictions.

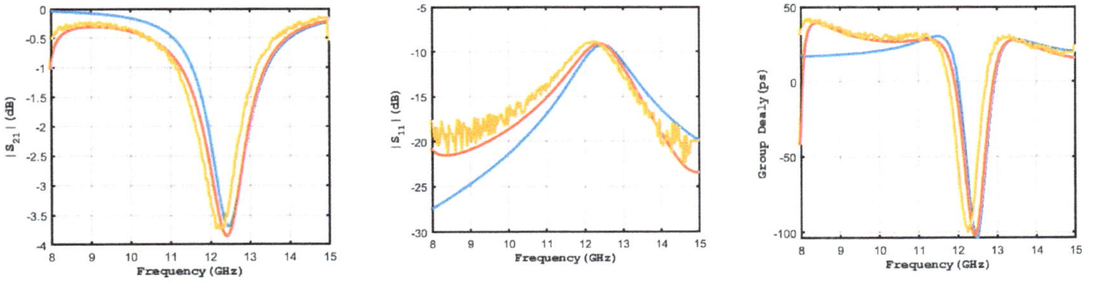

(a) (b) (c)

Figure 7. Measurements of the fabricated SR-SRR: (**a**) measured S21, (**b**) measured S11, (**c**) measured group delay.

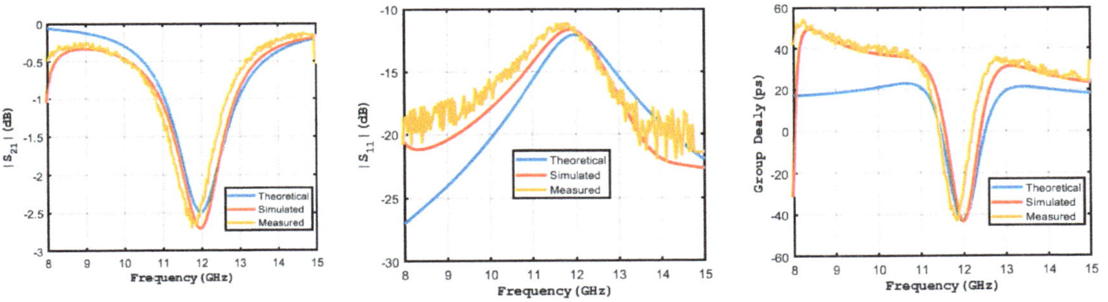

(a) (b) (c)

Figure 8. Measurements of the fabricated EC-SRR. (**a**) Measured S21, (**b**) measured S11, (**c**) measured group delay.

4. Application

The structure of continuous transverse stub (CTS) antenna is shown in Figure 9. The antenna comprises two layers: radiation and waveguide layers. The radiation layer is composed of transverse stubs. The electromagnetic power radiates from the slots between the stubs. The waveguide layer and the radiation layer form a parallel plate waveguide structure and antenna aperture.

The beam direction of CTS antenna with θ can be detected by

$$k\sin(\theta)p + k\sqrt{\epsilon_g}p = 2\pi \tag{10}$$

where θ is the elevation angle, λ is the wavelength, ϵ_g is the effective dielectric constant, k is the propagation constant in free space and p is the slot period.

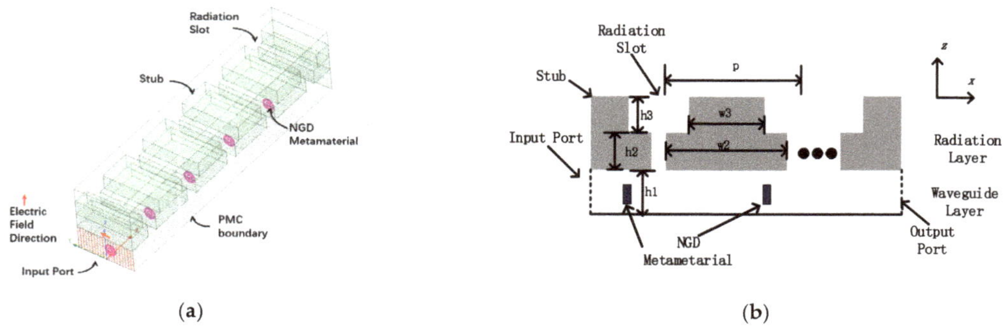

Figure 9. Diagrams of CTS antenna: (**a**) three-dimensional perspective and (**b**) side view.

It can be seen from Equation (10) that as the frequency increases the beam direction moves closer to the normal direction. The frequency scanning is caused by the time delay between each slot. Therefore, the NGD can be introduced to reduce the frequency scanning. NGD metamaterial can be placed in the waveguide layer or below the radiation layer, but here only the waveguide situation is discussed and EC-SRR was used. It should be noted that the metamaterial element period should be same as the slot period and the NGD value and frequency should be adjusted using the metamaterial parameters.

Simulations of the proposed CTS antenna were conducted using the Ansys HFSS suite (Ansys Inc., Canonsburg, PA, USA). The boundaries of the two sides are set using the perfect magnetic conductor (PMC), which are the same as for the CTS antenna, and the TEM mode is generated in the waveguide. For the CTS antenna, the TEM mode feed can be achieved by linear source generator. CTS antenna with five slots was simulated. The parameters were h1 = 8 mm, h2 = 5 mm, h3 = 5 mm, p = 19 mm, h2 = 5 mm, w2 = 15 mm, w3 = 11 mm, r = 1.5 mm, t = 0.5 mm, w = 0.5 mm, h = 0.1 mm, and d = 0.5 mm.

The simulated E-plane patterns of the CTS antenna are shown in Figure 10. The beam walk reduction can be observed. The simulated S parameters are shown in Figure 11. S_{11} and S_{21} are both below −12 dB.

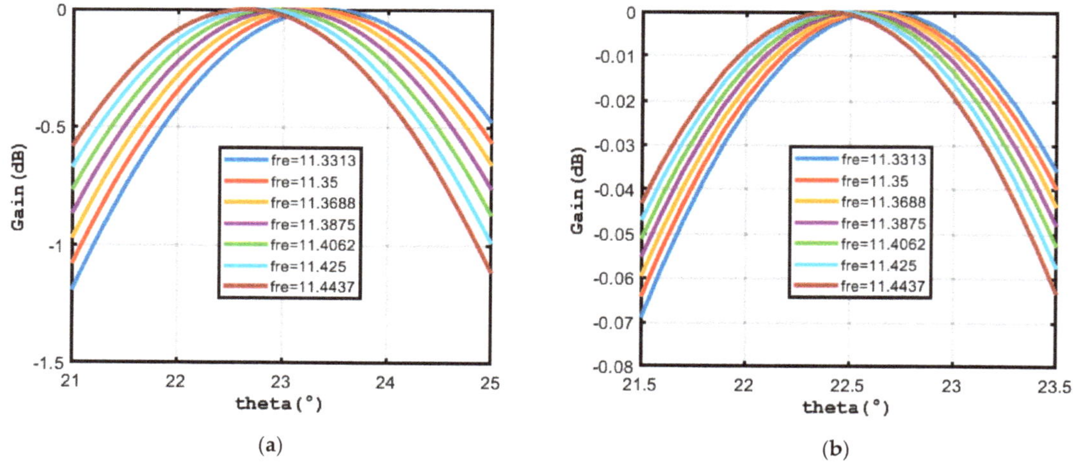

Figure 10. Simulated pattern of CTS antenna (**a**) without SRRs, (**b**) with SRRs.

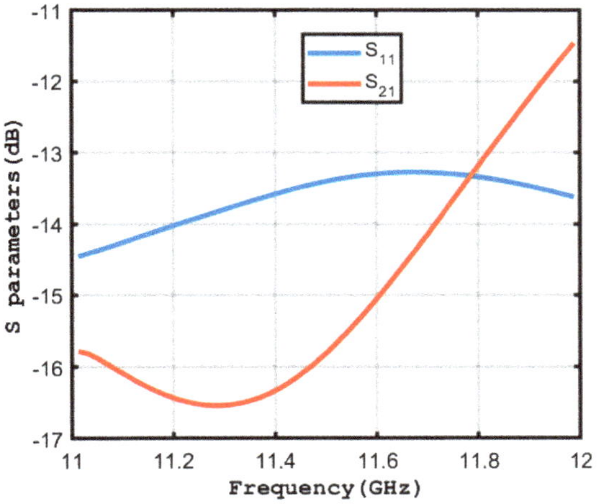

Figure 11. Simulated S parameters.

It should be noted that although the frequency scanning can be reduced with the introduction of the NGD metamaterial, the gain of the antenna also reduced. Additionally, because the gain varies with frequency, the instantaneous bandwidth may also be reduced. However, this issue can be resolved using the wideband flat NGD device.

5. Conclusions

In this report, negative group delay (NGD) metamaterials based on split-ring resonators (SRRs) were evaluated. A theoretical analysis was first performed to obtain the equivalent circuit parameters, NGD values, and S_{21} amplitudes. The parameters were then simulated and discussed, and the metamaterials were fabricated from tantalum nitride before being tested in a laboratory setting. The measured results were found to be consistent with the theoretical and simulated predictions. For EC-SRR, a negative group delay of up to -0.1 ns was achieved at 12–13 GHz. For SR-SRR of the same size as the out ring of EC-SRR, a negative group delay of up to -0.04 ns was achieved with a loss lower than 2.7 dB. The application of the metamaterial was discussed, and the frequency scanning reduction in CTS antenna was achieved using NGD metamaterials.

Author Contributions: Conceptualization, Z.L. and J.Z.; methodology, X.L.; software, J.G.; validation, T.L.; formal analysis, Z.X.; investigation, Z.L.; resources, Z.L.; data curation, Z.L.; writing—original draft preparation, Z.L.; writing—review and editing, Z.L.; visualization, Z.L.; supervision, Z.L.; project administration, Z.L. All authors have read and agreed to the published version of the manuscript.

Funding: The authors have no funding sources to declare.

Data Availability Statement: The data presented in this study are available on request from the corresponding author.

Conflicts of Interest: The authors declare no conflict of interest.

References

1. Pendry, J.B.; Holden, A.J.; Robbins, D.J.; Stewart, W.J. Magnetism from conductors and enhanced nonlinear phenomena. *IEEE Trans. Microw. Theory Tech.* **1999**, *47*, 2075–2084. [CrossRef]
2. Chen, Z.; Tang, M.-C.; Li, M.; Su, M. Far-Field Decoupling of Two-Element Antenna Transceiving System by the Periodic Near-Field Resonators. *IEEE Antennas Wirel. Propag. Lett.* **2022**, *21*, 2065–2069. [CrossRef]
3. Huang, D.; Xu, G.; Wu, J.; Wang, W.; Yang, L.; Huang, Z.-X.; Wu, X.-L.; Yin, W.-Y. A Microstrip Dual-Split-Ring Antenna Array for 5G Millimeter-Wave Dual-Band Applications. *IEEE Antennas Wirel. Propag. Lett.* **2022**, *21*, 2025–2029. [CrossRef]
4. Monti, G.; Tarricone, L. Negative Group Velocity In A Split Ring Resonator-Coupled Microstrip Line. *Prog. Electromagn. Res.* **2009**, *94*, 33–47. [CrossRef]

5. Woodley, J.F.; Mojahedi, M. Negative group velocity and group delay in left-handed media. *Phys. Rev. E* **2004**, *70*, 046603. [CrossRef] [PubMed]
6. Das, P.; Mandal, K.; Lalbakhsh, A. Beamsteering of microstrip antenna using single-layer FSS based phase-shifting surface. *Int. J. RF Microw. Comput.-Aided Eng.* **2022**, *32*, e23033. [CrossRef]
7. Liang, Y.; Koshelev, K.; Zhang, F.; Lin, H.; Lin, S.; Wu, J.; Jia, B.; Kivshar, Y. Bound States in the Continuum in Anisotropic Plasmonic Metasurfaces. *Nano Lett.* **2020**, *20*, 6351–6356. [CrossRef] [PubMed]
8. Lalbakhsh, A.; Afzal, M.U.; Esselle, K.P.; Smith, S.L. All-Metal Wideband Frequency-Selective Surface Bandpass Filter for TE and TM Polarizations. *IEEE Trans. Antennas Propag.* **2022**, *70*, 2790–2800. [CrossRef]
9. Brillouin, L.; Chako, N. *Wave Propagation and Group Velocity*; Academic press: New York, NY, USA, 1960. [CrossRef]
10. Takeda, S.; Anada, T. Phase equalizer making use of negative group delay times by reflection coefficients. In Proceedings of the 41st European Mi-crowave Conference 2011, Manchester, UK, 10–13 October 2011; pp. 627–630.
11. Chaudhary, G.; Park, J.; Wang, Q.; Jeong, Y. A design of unequal power divider with positive and negative group delays. In Proceedings of the 2015 European Microwave Conference (EuMC), Paris, France, 7–10 September 2015; pp. 127–130. [CrossRef]
12. Ahn, K.-P.; Ishikawa, R.; Honjo, K. Group Delay Equalized UWB InGaP/GaAs HBT MMIC Amplifier Using Negative Group Delay Circuits. *IEEE Trans. Microw. Theory Tech.* **2009**, *57*, 2139–2147. [CrossRef]
13. Froncisz, W.; Hyde, J.S. The loop-gap resonator: A new microwave lumped circuit ESR sample structure. *J. Magn. Reson.* **1982**, *47*, 515–521. [CrossRef]
14. Roshani, S.; Azizian, J.; Roshani, S.; Jamshidi, M.; Parandin, F. Design of a miniaturized branch line microstrip coupler with a simple structure using artificial neural network. *Frequenz* **2022**, *76*, 255–263. [CrossRef]

electronics

MDPI

Review

A Critical Examination of the Beam-Squinting Effect in Broadband Mobile Communication: Review Paper

Mariam Q. Abdalrazak [1], Asmaa H. Majeed [1] and Raed A. Abd-Alhameed [2,3,*]

[1] College of Information Engineering, Department of Information and Communication Engineering, Al-Nahrain University, Baghdad 00964, Iraq
[2] Faculty of Engineering and Informatics, University of Bradford, Bradford BD7 1DP, UK
[3] Department of Information and Communication Engineering, Basrah University College of Science and Technology, Basrah 61004, Iraq
* Correspondence: r.a.a.abd@bradford.ac.uk

Abstract: Due to the fact that the quality of the received signal is adversely affected by the beam squint phenomenon, which is especially pertinent to the millimeter wave (mmwave) bands, many studies have been utilised by other researchers to provide light on some of the challenges that come with analysing this type of occurrence. Squint causes several issues, the most important of which are its detrimental effects on gain, line of sight, angle of arrival, progressive phase, usable bandwidth, and fading effect. As a result of these obstacles, the advantages of adopting a high-frequency band such as mmwave in modern wireless communication systems are severely limited. Squint-phenomena-related difficulties, such as decreased channel capacity, increased bit error rate (BER), and lowered quality of services, may have a substantial detrimental impact on channel performance. The squint phenomenon and associated issues become more pressing with the expansion of the frequency spectrum and the deviation of the arrival angle from boresight. The purpose of this article is to provide a comprehensive overview of the relevant literature and to compare and contrast various methods in order to identify the most fruitful lines of inquiry for future research.

Keywords: beam squinting; massive MIMO; millimeter waves

Citation: Abdalrazak, M.Q.; Majeed, A.H.; Abd-Alhameed, R.A. A Critical Examination of the Beam-Squinting Effect in Broadband Mobile Communication: Review Paper. *Electronics* **2023**, *12*, 400. https://doi.org/10.3390/electronics12020400

Academic Editor: Luis Castedo

Received: 2 November 2022
Revised: 27 December 2022
Accepted: 5 January 2023
Published: 12 January 2023

1. Introduction

In today's world, almost everyone has a mobile device, and the number of people who do is growing. Concurrently, the number of interactive multimedia apps is also growing, and experts predict that this pattern will continue into the foreseeable future [1]. In spite of the fact that mobile technologies of the first, second, third, and fourth generations have spent the last 25 years working to increase the speed and efficiency of wireless networks, there are still some specific application areas in which current wireless networks are unable to deliver [2]. As the need for mobile traffic steadily develops, a bottleneck between spectrum constraints and capacity requirements is becoming more obvious. This bottleneck is becoming more and more apparent. The wireless bandwidth bottleneck is one of the most serious challenges faced by 5G communications [3]. Wireless carriers need to be ready to sustain a growth in total mobile traffic that may be as much as 1000 times higher than it is now since the need for capacity in mobile broadband communications is growing at a substantial rate each year. The pursuit of inventive solutions is required in order to make the transition to the fifth generation (5G) of wireless technology [4]. Because of the growing need for bandwidth, the millimetre wave spectrum has been standardised for use in high-speed wireless communication [5]. mmwaves are an interesting possibility since they have a significant amount of bandwidth and spectrum that is readily available (approximately 60 GHz) [6]. mmwave communications may make use of the large capacity that is now being underutilised, which will help them fulfil the high demands of the future generation of wireless networks [7].

Although the potential bandwidth of mmwave frequencies is promising, the propagation characteristics of these frequencies are very different from those of microwave frequency bands [8]. These propagation characteristics include path loss, diffraction and blockage, rain attenuation, air absorption, and foliage loss behaviours. The massive multiple-input multiple-output (MIMO) technology is a potential answer to this problem. With this technology, a considerable number of antennas might provide sufficient gain to compensate for the route loss [9,10]. Massive MIMO over mmwaves is a novel approach that combines the benefits of massive MIMO antenna arrays with the vast amount of bandwidth that is currently accessible over mmwaves [11]. Massive MIMO over mmwaves has the potential to revolutionise wireless communication. Beam squint is one of the issues that arise in mmwave communications. This issue manifests itself as a change in the spatial direction of a beam as a function of frequency. As a consequence, there are discernible variations in the path phases at various frequencies.

In this paper, we explore the research that has been conducted so far on the subject of beam squint and how it might be alleviated while also minimising the complexity and expense of hardware implementation. Specifically, we examine how it might be possible to fix the problem while also keeping the implementation as simple and inexpensive as possible. Additionally, the segment of the research that relates to the exploitation of beam squint is presented in this study. The following is the structure of the paper: In Section 2, we discuss the many issues that pertain to the broadband communication. In the third section, the research that pertains to the beam squint effects are presented. While the research concerning the beam squint exploitation are discussed in Section 4, this section also focuses on other topics. Section 5 is where the conclusion of the paper can be found.

2. Broadband Communication

2.1. Millimeter Waves and Massive MIMO

In order to attain more bandwidth than traditional frequency bands, 5G and the following 6G mobile networks operate in high-frequency zones [12]. This allows for high-capacity wireless transmission of data at speeds up to several gigabits per second (Gbps) [13]. It is generally agreed that mmwave communication [14] is one of the most crucial technologies for reaching peak data rates of 10 Gbit/s [15]. This is due to the fact that increasing the bandwidth is an effective method for increasing the system's capacity [7] and that there is a lot of bandwidth available in the mmwave channels. Measurement and modeling of mmwave channels [16] and assessment of the band's potential for future wireless systems have received significant academic and industrial attention because of the spectrum's abundance [17]. However, due to negative propagation effects, particularly those brought on by obstructions in the line of sight (LoS), these high frequencies have been used only sparingly up to this point [18]. It is important to handle the greater propagation losses associated with mmwave frequencies, especially beyond 100 m and in non-line of sight (NLoS) environments [19]. On the one hand, signal transmission becomes more difficult in areas with dense foliage and higher propagation loss [20]. Nonetheless, the link could be forced into the low signal-to-noise ratio (SNR) zone due to the constrained send signal power dispersed over a broad bandwidth [21]. To address these problems, several transceiver topologies have been created that steer the received or broadcast beams in the appropriate direction. [22]. The two main technologies that will enable 5G mmwave communication are MIMO and beamforming [23]. Beamforming is a method for using MIMO/array elements to direct the main lobe radiation beam at the transmitters and receivers in the desired directions, eliminating the undesirable spatial selectivity [24]. The beamforming array is widely divided into the following categories: analog, digital, and hybrid beamforming array [25]. Figure 1 shows the block diagram of massive MIMO system.

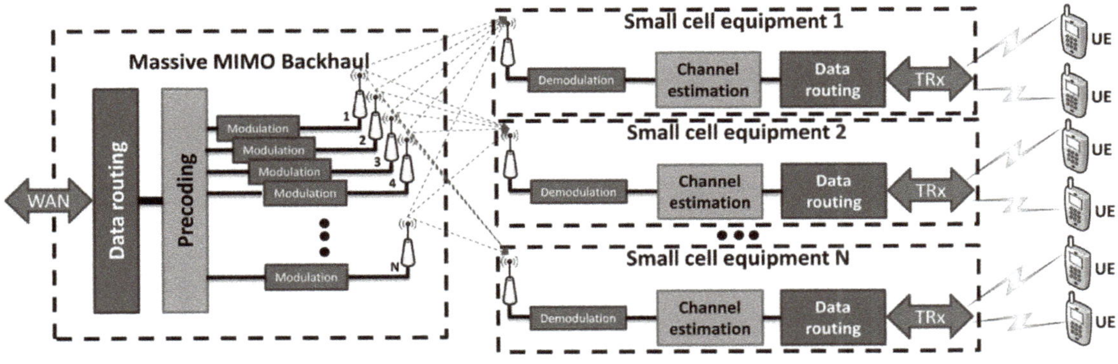

Figure 1. Functional block diagram of massive MIMO system [26].

2.1.1. Analog Beamforming

The architecture of analog beamforming is shown in Figure 2. A phased array is used to accomplish analog beamforming, and the single RF chain is powered by either an analog-to-digital converter (ADC) in the receiver or a digital-to-analog converter (DAC) in the transmitter. Frequency-up converters, power amplifiers, and other components form a transmitter's RF chain; low-noise amplifiers, frequency-down converters, and other components form a receiver's RF chain [27].

The phased array's antenna weights are required to be digitally controllable phase changes. The relative amplitudes of the signal sent into the antennas of the transmitter cannot be adjusted, and the phases of the phase shifters are normally quantized to low resolution [28]. A beam is created as a result of the send signal, which is constructive in certain directions and destructive in others. On the basis of precise beam-steering tactics, the phase shifters' phases can be dynamically changed. The receiver has a similar set of features [27].

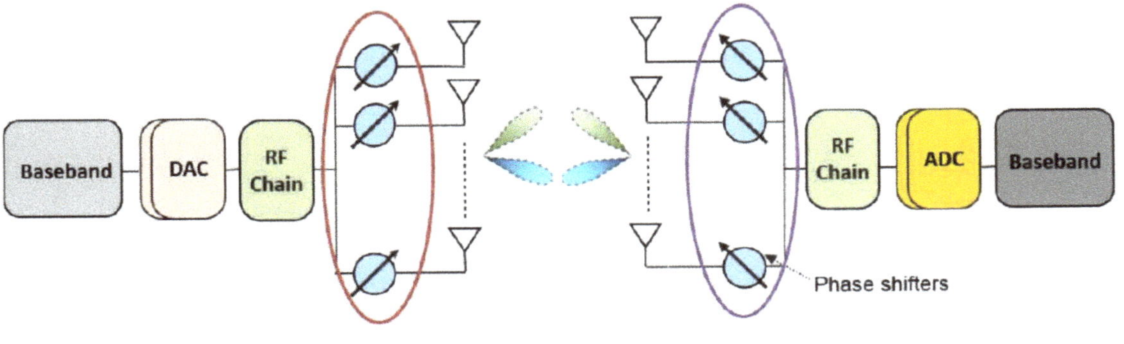

Figure 2. An mmwave MIMO system using analog beamforming [29].

2.1.2. Digital Beamforming

The radio frequency (RF) signal in a digital array is transformed into a digital signal at the sub-array or element level, and beamforming is accomplished using a digital signal processor [30]. In order to process the control signals for beamforming, a digital signal processor is used, which offers additional flexibility and degrees of freedom for the implementation of effective beamforming algorithms. Due to the method's requirement for a separate RF chain for each antenna element, it consumes a great deal of power and has a complicated architecture [31]. Figure 3 shows the structure of digital beamforming.

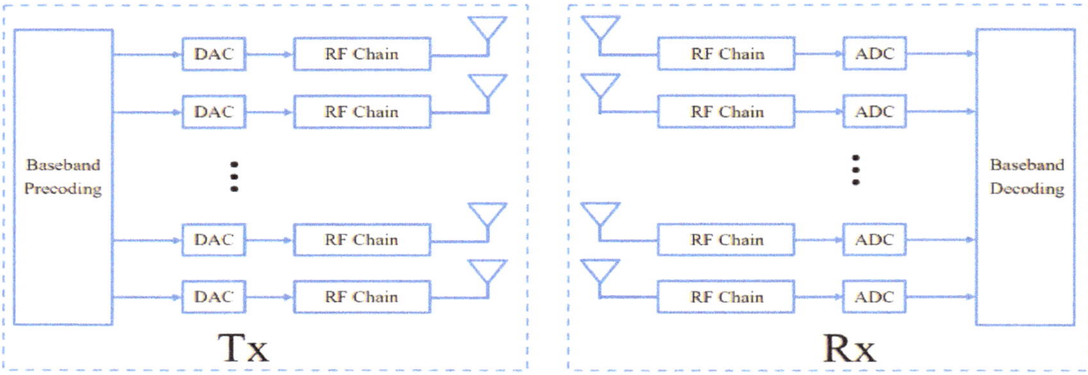

Figure 3. Digital beamforming [31].

2.1.3. Hybrid Beamforming

Recently, interest in the hybrid analog/digital beamforming array has increased [32,33]. This type of array increases the efficiency of analog beamforming while reducing the complexity of digital beamforming [34]. The hybrid beamformer is composed of two parts: an analog and a digital component [35]. The RF chains that compose the digital portion share a common structure with the many plans under discussion. For each user on each subcarrier, a hybrid beamforming's digital component can be used, just like in traditional fully digital beamforming. In contrast, this is not true of the hybrid beamforming analog network or analog component. Since the transmitted signals for all users are combined by the digital beamformers and analog RF beamforming is a post-IFFT (inverse Fast Fourier Transform) operation, all users and subcarriers in an analog network share the same RF component [36]. Hybrid beamforming's fundamental architecture is depicted in Figure 4.

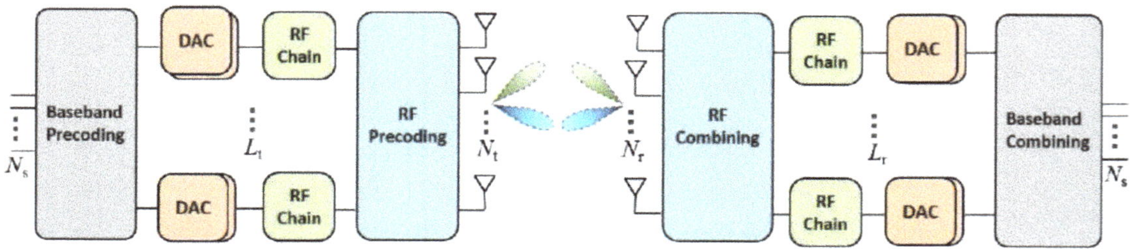

Figure 4. Hybrid beamforming [29].

Beam squint is a phenomenon caused by the frequency dependence of the magnitude of the phase shift in analog beamforming [37]. This can cause issues with wideband communication. The Digital Signal Processing (DSP) logic and digital beamforming algorithms can both use actual time delay. As a result, a phased array architecture with every element digitalized would naturally lend itself to solving the frequency dependency problem as well as offer the greatest degree of programmable flexibility. However, issues with this solution's strength, size, and price can arise.

A time delay among elements occurs as a wavefront approaches an array of elements, and it depends on the wavefront angle in relation to boresight. The beam steering can be achieved with a single frequency by substituting a phase shift for the time delay [38]. This is true for narrowband waveforms, but in wideband waveforms when a phase shift is used to provide beam steering, the beam's direction can change with frequency. As a result, the needed phase shift varies with frequency for a given beam direction. Alternately, the beam direction varies with frequency for a given phase shift. Beam squint is the effect where the beam angle changes depending on the frequency [39]. Figures 5 and 6 shows the effect of beam squint.

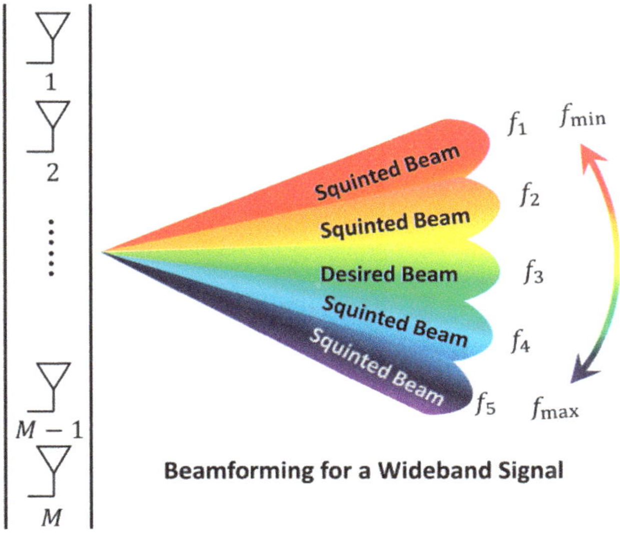

Figure 5. The effect of beam squint [9].

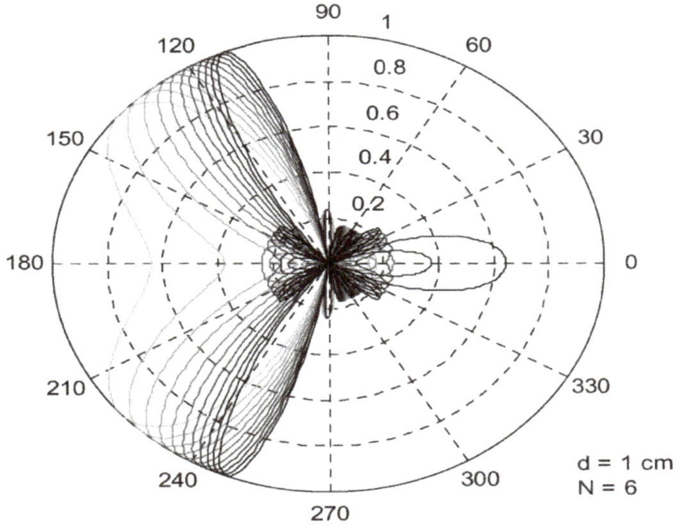

Figure 6. Beam squint effect for a phased array antenna using electrical phase shifters operating at frequencies in the range of 10–20 GHz [39].

In addition, keep in mind that there is no phase shift between the elements at boresight ($\theta = 0°$); hence, there is no way to induce any beam squint. As a result, both the frequency variation and the angle must be functions of the amount of beam squint [38]. Beam squint can seriously affect performance since mmwave communications depend heavily on accurate beam alignment between the transmitter and the receiver [40].

2.2. Effect of Beam Squint on the Communication System Performance

- Bandwidth and Capacity:

The system bandwidth in wideband communication is constrained by the beam-squint phenomenon [41]. The beam squint effect is closely correlated to bandwidth; as bandwidth rises, the beam squint effect becomes more pronounced, and a larger fraction of the subcarriers experience tiny array gains. Even outside the main lobe, certain subcarriers may exist. As a result, the whole system's capacity starts to decline [42]. Figure 7 shows the channel capacity versus the bandwidth with and without beam squint. It is clear from this figure that the capacity decreases as the array's element count, fractional bandwidth, and beam focus angle rise.

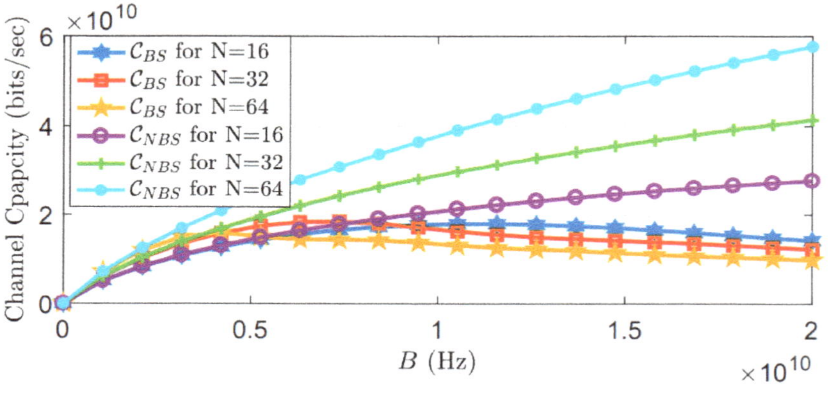

Figure 7. Capacity as a function of bandwidth with beam squint (CBS) and without beam squint (CNBS) [41].

- Channel estimation:

Channel estimation algorithms for Orthogonal Frequency Division Multiplexing (OFDM), Maximum Mean Square Error (MMSE) estimator, and Maximum Likelihood Estimation (MLE) estimator are widely used [41]; however, these estimators do not consider beam squint phenomena. In [42], the MLE channel estimator was used to investigate the effect of beam squint on the estimation process, Fading was considered fixed, and the results showed that beam squint increases the error of channel estimation, even without any noise. The error also increases with increasing the number of antenna elements in the array, fractional bandwidth, and the magnitude of beam focus angle. Figure 8 shows an example of the MLE channel estimation with beam squint, where H(n) represents the true channel, and $\acute{H}$(n) represents the estimated channel. It is clear from this figure that there is a difference between the actual channel and the estimated channel, even without considering the noise.

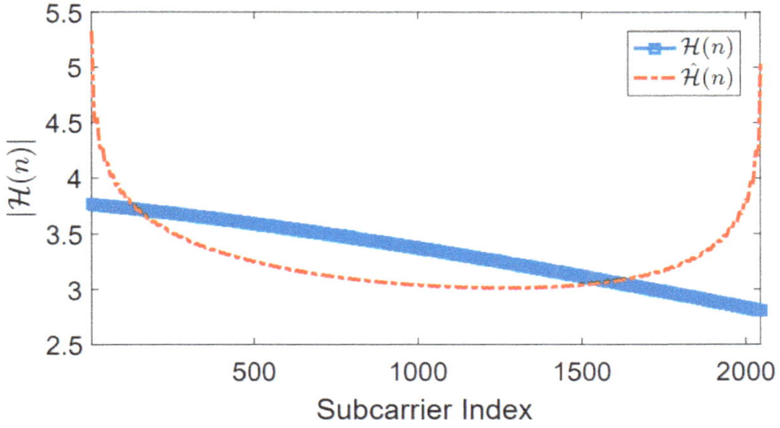

Figure 8. Example of MLE channel estimate using the receiver's beam squint. There is no beam squint in the transmitter and only one path. No more noise exists. The number of antennas N = 16 [41].

3. Studies Related to the Beam Squint Effects

Since the 1990s, beam squint phenomena in antenna attracted much attention; in [43], the authors drew a comparison between prime focus beam squint and offset fed of microstrip reflect arrays. Results from an examination of a linear reflectarray are presented, and a straightforward criterion for decreasing beam squint with frequency is discussed.

The author in [44] describes the design of a tri-mode matched feed horn in order to counteract the effects of beam squinting in a circularly polarized offset parabolic reflector antenna. Three modes—TE11, TM11, and TE21—are merged in a conical horn at the correct amplitude and phase ratio to provide a tri-mode matched feed arrangement. The circularly polarized offset parabolic reflector antenna is then lit using the suggested tri-mode horn as the primary feed mechanism. The radiation parameters of the offset reflector are simulated, the amount of beam squinting is calculated, and the results are compared with those of a conventional offset reflector fed by a potter horn. The study also includes the experimental findings regarding the secondary radiation pattern.

In [45], in order to calculate beam squinting in a phased array antenna system, a general formula was created. This formula is especially useful for estimating wide band beam squinting (time-delay) in phased array antenna systems. To determine beam squinting, the authors calculated the criterion (fϕ = 0, cell) from the delay cell's phase-transfer function, then used (fϕ = 0, cell) as a parameter in the beam-squinting formula.

In radar communications, the beam squint phenomena were studied in [46]. By employing phase shifters as opposed to time-delay, Matt Longbrake was able to construct a formula for beam squint and demonstrated how the array factor varies. In order to address the beam squint issue, true temporal delay (TTD) beam steering was applied using two different techniques, the first of which was dubbed optical delay lines, and the second, electronic delay lines. In the first technique, an RF signal was used to modulate a laser diode's bias current. A length of optical fiber delays the light, which is subsequently transformed by a photodetector back into an electronic signal.

The main drawback of optical time delay was the modulator's and detector's subpar RF performances, particularly insertion loss. Traditional microstrip lines or coax wire was employed in electronic technologies to delay the signal. Overall, the results demonstrated that TTD beam steering outperforms phase shifters; however, this implementation came at a considerable cost and complexity of the system.

Waleed A. and Amir M. in [47] implemented a novel design for a serially fed antenna array using a negative group delay (NGD) in their study of the beam squint problem in mobile communication [48]. Each active antenna unit included an amplifier, a T-junction power divider, and an antenna in the array of series-fed active antennas that composed the design (see Figure 9).

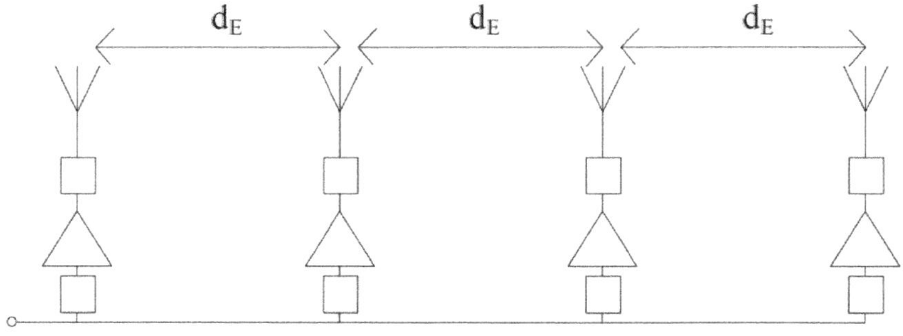

Figure 9. Series-fed active antenna array [47].

Beam squinting was a significant problem with the conventional series-fed antenna array. To solve this issue, the phase shift among the antennas should be frequency-independent. In other words, there cannot be a delay in the group. To achieve zero-group delay between the adjacent antennas, a negative group delay (NGD) circuit was added to the feed line. The variance of phase shift with frequency was decreased and the beam squint was minimized by fusing the NGD circuit with the connecting line. With no additional losses, this method significantly reduced the beam squint problem by a factor of six.

In [49], the authors investigated the 60 GHz mmwave wireless communication beamforming process's beam squint issues in 2013. The goal of beamforming was to choose the best weight vectors for the transmitter and reception antennas with high antenna gains or the best beam pattern pairings, and this criterion depended on several factors, including capacity, the signal-to-noise and interference ratio (SNIR), and others. It was found that the phenomena of beam squint are caused by distinct frequency bands away from the 60 GHz core frequency.

The authors established three different codebook strategies in order to cut down on beam squint. The initial methodology is referred to as the "Phase Improvement Scheme Based on 3C Codebook." The authors of this plan developed codebooks for a number of different bands and altered the phase by utilising an additional phase shifter. The suppression of the beam squint resulted in a reduction in the amount of spatial interference and an increase in system capacity. On the other hand, the enhanced system capacity came at the expense of an increase in the complexity of the implementation. A comparison of the most common approaches of removing the squint effect from beams is shown in Table 1.

Table 1. Comparison among the main methods used to remove beam squint effect.

Parameter	Digital Beamforming	Analog Beamforming	Codebook Design
Complexity	High complexity	Acceptable complexity	Increases the number of phase shifters, which increases the complexity
Cost	Increased hardware requirements lead to a high cost	Compared with their digital counterparts, the costs are much lower	Power and cost rose significantly
Effectiveness	Incredibly efficient at correcting the squint	Has the potential to decrease the effect of squinting and improve efficiency	Able to significantly lessen the "squint" effect

The authors proposed the "MRA Alignment Scheme Based on Uniform-Weighting Codebook" as the second scheme to enhance system performance. The final proposal assumed that since neighboring parts have a strong correlation, it was possible to alter the phase of two nearby antennas using a single-phase shifter, which significantly reduced the complexity of the transceiver.

Figure 10 shows compression among the three schemes as a function of the system capacity. Although these schemes worked very effectively in removing the beam squint, they increased the system complexity by increasing the number of phase shifters.

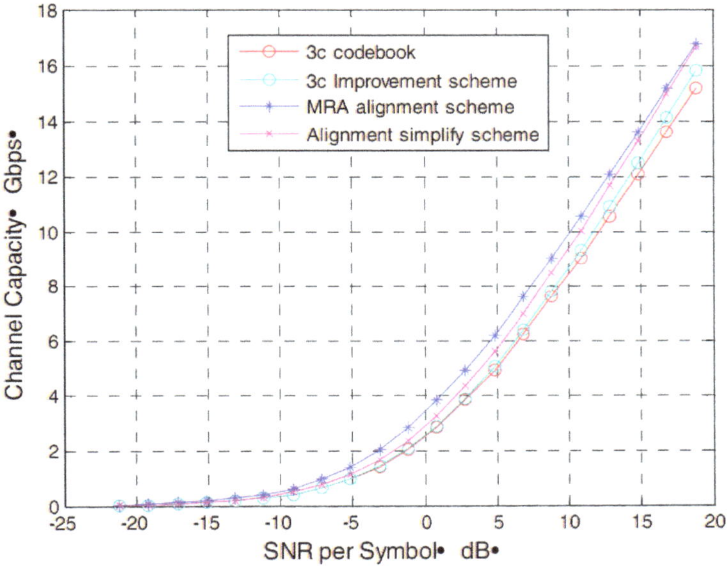

Figure 10. Point-to-point capacity analysis [49].

In [50], the beam squint problem in Uniform Linear Array (ULA) was modelled, and its effect on codebook design was analyzed. The study demonstrated that beam squint reduces the bandwidth that is usable. The authors created an algorithm to fix codebook designs with beam squint. The goal of codebook design was to ensure that every beam exceeds a specified threshold and has a minimal gain for a variety of frequencies and angles in the wideband system. Analysis and numerical examples indicated that to account for beam squint, a denser codebook was needed. In other words, compared with a codebook design that disregards beam squint, more beams were required.

In order to boost the expected average beam gain, the authors of [51] studied the beam squint issue in mmwave communication and suggested a beamforming architecture that optimizes average beam gain inside the bandwidth while minimizing average beam gain outside the bandwidth. Additionally, the design shows little variation in the bandwidth at various frequencies. To take into account the beam squint and guarantee constant beam gain for each subcarrier in the wideband system, the authors suggested a beamforming approach based on space–time block coding. In order to offer a less-than-ideal answer, on the basis of eigenvalue decomposition, the semidefinite relaxation (SDR) approach was applied. The simulation results showed that, under specific circumstances, the suggested transmission method can effectively reduce beam squint and improve wideband communication systems' throughput performance.

In order to demonstrate how the number of antenna elements influences beam squint and how wideband beam squint affects the traditional narrowband models, an analysis of capacity was carried out in [52]. Combining two state-of-the-art cooperative algorithms led to the suggestion of a method for assigning subcarriers to beams. The first methodology determines the optimal beams for allocation in a multi-carrier implementation and determines how much beam squint is present in the system, whereas the second methodology maps data to the appropriate beams. The results of the investigation showed that beam squint significantly affected system capacity. The results also showed that beam squint might be utilized to better manage resources and expand system capacity, as shown in Figure 11.

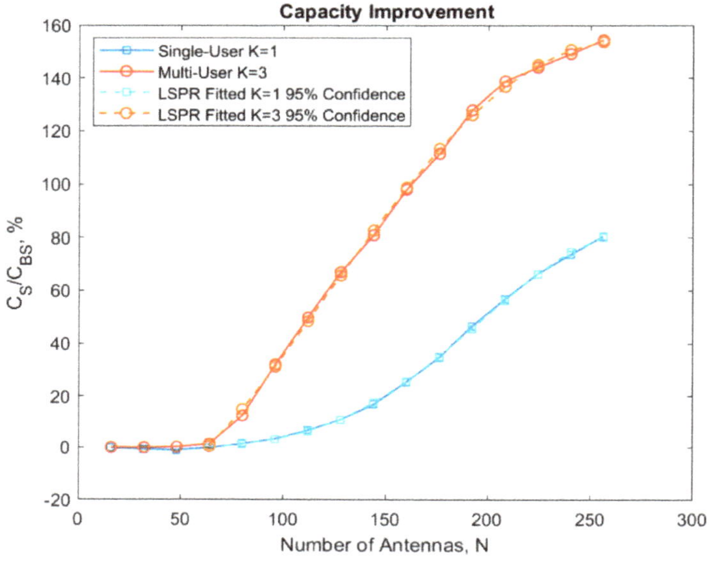

Figure 11. Improvement of capacity for fc = 73 GHz, BW = 2.5 GHz [52].

The beam squint effect on channel estimation had been analyzed in [53]. In this study the authors clarified that beam squint in an MIMO-OFDM system causes the subcarriers to see distinct angles of the same path and renders the existing channel estimation and precoding algorithms inapplicable. For frequency-division duplex (FDD) mmwave massive MIMO-OFDM systems with hybrid analog/digital precoding, the authors suggested a channel estimation approach that takes the beam squint effect into account. The angle of arrival (AoA) and time delay, which are frequency-insensitive parameters of each uplink channel path, as well as the complex channel gain, which is frequency-sensitive, were extracted using a compressive sensing-based methodology. The reciprocity of these frequency-insensitive parameters in frequency division duplex (FDD) systems allows for a significant simplification of the downlink channel estimating process, using only a small number of pilots to acquire the downlink complex gains and reconstruct the downlink channels. The results of this study showed that the proposed channel estimation scheme can perform better than the other conventional methods under general system configurations in mmwave communication.

The authors of [54] developed an off-grid CSS technique to calculate the uplink channel and offered a non-negligible beam squint in the field of channel estimation with beam squint. According to the authors, the direction of arrival (DOA) and the delay parameters might be calculated from the signals received on the uplink channel's pilot subcarriers using a shift-invariant block-iterative gradient technique. The entire set of channels over all subcarriers was then rebuilt using the DOA and delay settings. The outcomes demonstrated that this offered superior outcomes in comparison with previous approaches that neglected to account for beam squint.

The authors in [55] present a unique digital method for correction of beam squint by utilizing Nonlinear Frequency Modulation (NLFM) Tansec waveforms when addressing phase-based phased array antennas. An algorithm was designed that chooses Tansec waveforms with phases that rectify FM-induced phase variations in the necessary transmitted signals phases. Simulations for a typical 77 GHz automobile radar with a 3 GHz BW show that the suggested technique can reduce beam squint.

The authors of [56] examined the issue of estimation of channel for mmwave MIMO systems with hybrid beamforming. A beam squint for channel estimation and an asymptotically ideal analog-beam scanning and pilot tone allocation system were developed. The suggested architecture is expanded to include the Uniform Planar Array (UPA) situation that satisfies the channel vectors' asymptotic orthogonality. To further support the effectiveness of the proposed design, a numerical comparison with the computed Cramer Rao lower bound (CRLB) was performed.

The wideband mmwave communication supported by a reconfigurable intelligent surface (RIS) suffering from the beam-squinting effect was first studied by [57]. Each component of the RIS is a nearly passive device, similar to the phase shifter [58], and it is a novel sort of "array" that has garnered considerable interest from both academics and the industry. Compared with a conventional antenna array, which has energy-hungry radio frequency chains, RIS is able to adaptively reflect the incident signals in the desired directions while using less power. Given that a passive reconfigurable intelligent surface (RIS) is applied in time domain and that mmwaves communications will have bandwidths up to several GHz, the phase shifts are assumed symmetrical for all frequencies. This will result in significant performance degradation. Several phase shifter design ideas were put forth to address the beam squint issue, taking into consideration both line of sight and non-line of sight instances.

A near-optimal phase shift design approach was developed, which is dependent only on the long-term angle data after identifying the best phase shift for each frequency in the LoS scenario. To achieve this, the obtained upper bound of the cumulative attainable rate was maximized. For the non-line of sight (NLoS) scenario, a method based on the mean channel covariance matrix (MCCM), was presented in order to determine the common phase shift for all frequencies. When a large number of components or a high bandwidth were used in the RIS, the detailed numerical analyses showed that the beam squint would result in a performance loss of more than 3 bps/Hz, and the effectiveness of the suggested approaches was proven to lessen the influence of the beam squint.

In [59], two wideband hybrid beamforming techniques were proposed. The authors' first approach suggested dividing the array into virtual subarrays, as illustrated in Figure 12, to produce a bigger beam and provide an evenly distributed array gain across the entire operational frequency spectrum. Results showed that this technique can reduce beam squint to some extent.

The second method modified the analog beamformer/combiner for a hybrid phased array transceiver on the basis of true time delay lines (TDD). Instead of replacing the phase shifters entirely, a small number of TTD lines are utilized to reduce the costs. The phase shifters were divided into groups and each group had one TTD line, as shown in Figure 13. The results of this method showed better results in removing beam squint while reducing the cost.

In [60], the authors investigated the impact of beam squinting on single carrier frequency domain equalization (SC-FDE) transmission systems used in mmwave communications. In high-throughput mmwave LoS multiple-input single-output (LoS-MISO) systems, the response of a uniform linear array (ULA) antenna and the common analog beamforming vector was described as a spatial equivalent channel. It was discovered through analysis of the spatial equivalent channel's characteristics that it behaves similarly to frequency-selective fading. The spatial equivalent channel's deep fading points were eliminated using an improved analog beamforming technique based on the Zadoff-Chu (ZC) sequence. The receiver then uses low-complexity linear zero-forcing and minimal mean squared error equalizers to reduce ISI brought on by the beam squint. The simulation results show that the proposed analog beamforming based on ZC can successfully reduce the performance impact of beam squint.

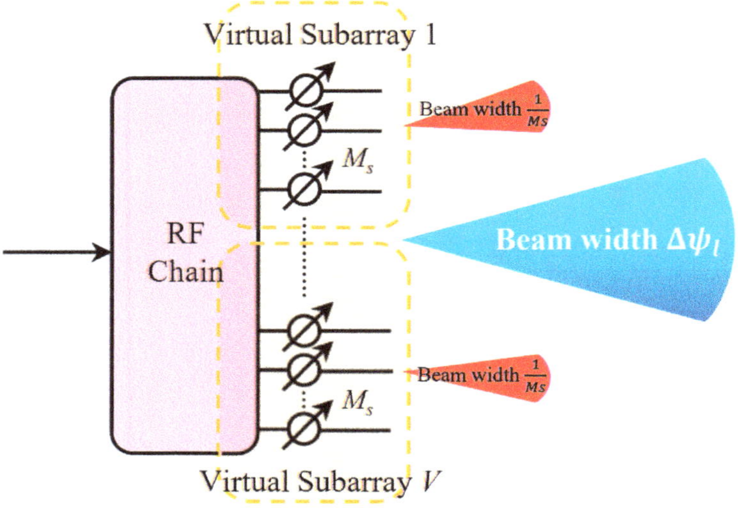

Figure 12. Diagram of the proposed subarray design [59].

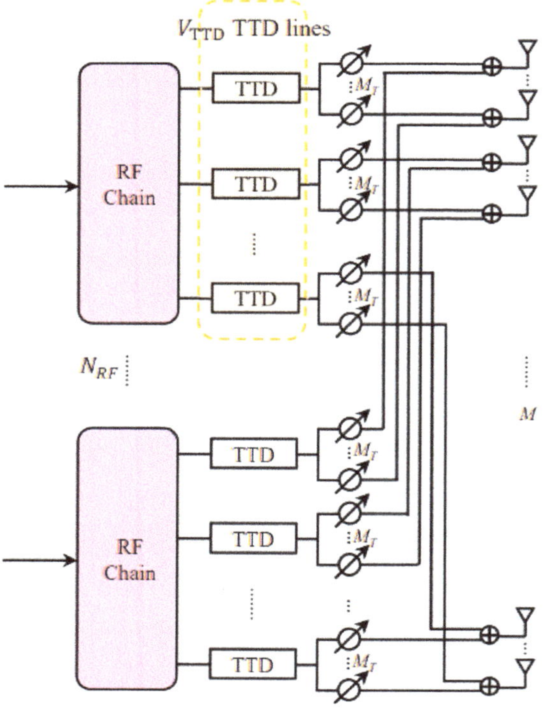

Figure 13. Diagram of the proposed analog beamforming design with true time delay lines [59].

Beam squint in RF lens antenna was discussed in [61] for ultra-wideband millimeter-wave (mmwave) systems. According to the authors of this study, phased array antennas are not affected by the same factors that cause RF lens antenna beam squint. According to Snell's law, if an electromagnetic wave passes through a dielectric lens, it will be bent. Refraction is a result of permittivity being a function of frequency, one of the reasons of beam squint, according to the Drude–Lorentz model [62]. Due to the resonance frequency and damping constant, permittivity is also nonlinear. Permittivity has both genuine and fictitious components. The real component is represented by the dielectric constant, and the tangent loss is the ratio of the real to the imaginary component. As can be seen, the component of the refractive index, known as the real part, is the one that determines the degree to which the beam is deflected. Because permittivity affects the actual portion of the refractive index, the beam squints at the lens when a wave with a different frequency band travels through a material with a different degree of refraction, as shown in Figure 14. This occurs because a wave with a different frequency band travels through a material with a different degree of refraction.

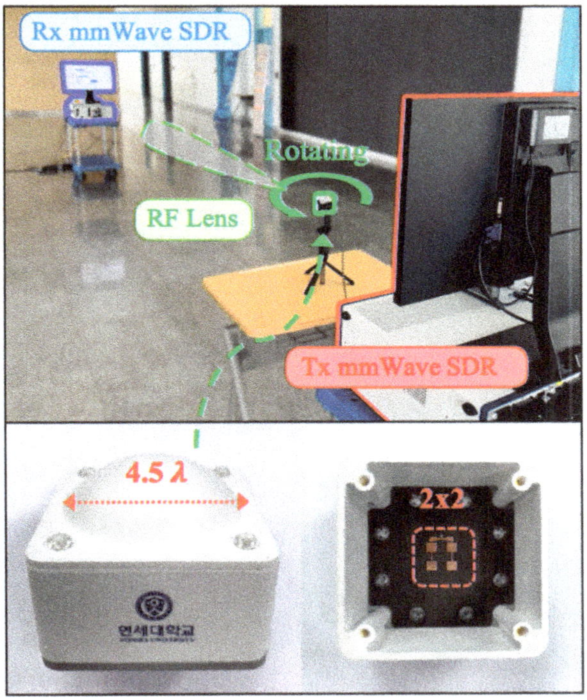

Figure 14. Indoor mmwave software-defined radio (SDR) testbed with a fabricated RF lens for beam squint [61].

To make fully digital beamforming less complicated, the authors developed hybrid beamforming. The lens structure at the front of the antenna array and the antenna array itself formed the two components of the RF lens antenna. The analog beamforming process uses the lens structure to concentrate the beam produced by the antenna array. With the antenna switched on and off, beam steering was accomplished utilizing lens refractions. Under comparable circumstances for beam squint, beamforming gain, and received power, the authors compared and contrasted the lens antenna and the phased array. Through study, they were able to confirm that stable permittivity materials can be used to minimize the beam squint issue with the RF lens system. Then, they demonstrated the reduced spectral efficiency brought on by beam squint for both the lens antenna and the phased array using 3D ray tracing in an interior setting. Finally, they used the constructed RF lens to illustrate the indoor mmwave link level and to confirm that beam squint negatively affects performance.

In [63], the authors proposed a transceiver design based on lens antenna subarray (LAS) and analog sub-band filters to compensate for the beam squinting. The design is based on the idea of dividing the ultra-wideband signal into narrowband beams and controlling them with a simplified exhaustive search-based precoding that is proposed to align the beam angle to the target direction. Analysis was based on various performance parameters, such as beam gain, complexity, power consumption, and capacity. The proposed design involved using both phase shifters and switching networks to steer the beam to compensate for the beam-squinting problem while maintaining the intended beamwidth performance. Phase shifters were used to steer the beam to the desired direction on the basis of the location of the targeted user, and the switching mechanism was used to select an antenna element under the lens that can correct/minimize the deviation of the beam (due to squinting) from the intended direction. The results of this study showed significant performance improvement. However, this study requires the use of high-quality narrowband filters for higher frequencies. This can increase the complexity of the system as well as the cost.

Another approach, based on reconfigurable intelligent surface for satellite-to-terrestrial relay networks, was developed and described in [64]. The Taylor expansion and penalty function methods were used in this approach to optimise phase shifters between the surface unit cells by adding a weight function in order to maximise the channel performance under certain conditions.

4. Studies Related to the Beam Squint Exploitation

Beam squint is not always a bad thing; the idea of the main beam changing direction with frequency was appealing for many researchers, and some studies even used it to enhance system performance. For instance, in [65], the authors formed many beams using the beam-squinting property of planar arrays using wideband perfect phase shifters instead of an existing hybrid structure. The following steps can be used to describe the work: initially, the authors determined the beam characteristics of the wideband array. Second, they created a coplanar array using the concept of the fractal that can operate across three mmwave frequency bands. The array's phase shifter was then swapped with one based on cells, as shown in Figure 15, which illustrates the steps for the suggested Sierpinski carpet antenna array with three antenna tiers, and in Figure 16, which illustrates the extensions and variations of the three-tier Sierpinski carpet array. Finally, they created the squinting multi-beam algorithm that facilitates multi-user communications on the basis of the beam-squinting property of the wideband array with optimal phase shifters. Under ideal circumstances, the study demonstrated an improvement in system performance (mutual coupling and isotropic elements were assumed).

In [66], a joint optimization design based on non-orthogonal multiple access-based satellite–terrestrial integrated network at mmwave was developed for cellular network technology. They developed, in their work, a user pairing scheme, therefore; non-orthogonal multiple access technique was exploited through grouping more than two users in the same cluster.

In [67], the authors explained how beam squint can be utilized for user localization. The author's idea was to control the beam squint using time delays and using it for localization. They created a method to regulate the trajectory of these beam squint points by deriving a trajectory equation for near-field beam squint points, as shown in Figure 17. In order for users in various positions to obtain the most power at various subcarriers, beamforming from various subcarriers would purposefully point to various angles and distances. Therefore, using the beam squint effect, one can easily determine the positions of the various users. The efficiency of beam squint in user localization can be shown from the study's findings. In addition, the same authors in [67] applied an advanced technique, called reconfigurable intelligent surface (RIS) [68], which suppressed the beam-squinting effect in broadband mobile communication.

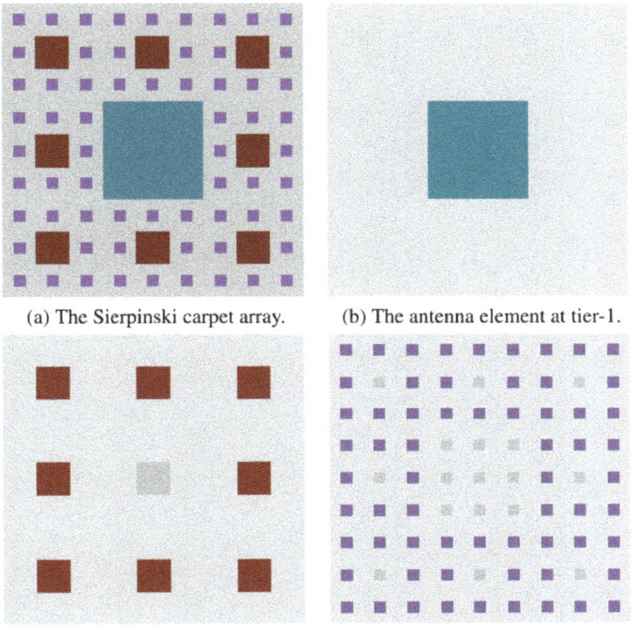

(a) The Sierpinski carpet array. (b) The antenna element at tier-1. (c) The antenna array at tier-2. (d) The antenna array at tier-3.

Figure 15. A Sierpinski carpet antenna array with three antenna tiers [65].

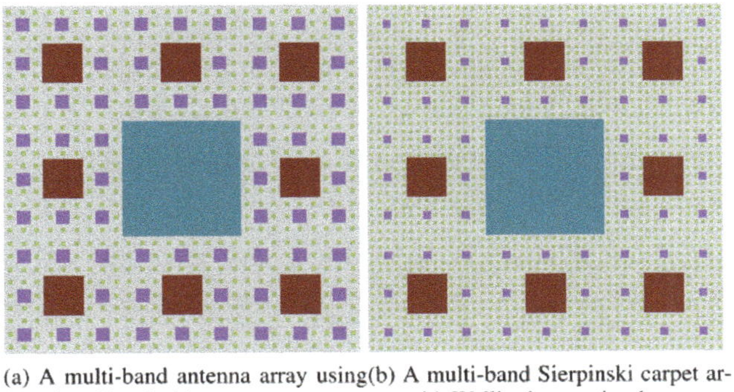

(a) A multi-band antenna array using four tiers Sierpinski carpet. (b) A multi-band Sierpinski carpet array with Wallis sieve at tier-4.

Figure 16. Extensions and variations of the three-tier Sierpinski carpet array [65].

We believe that, in the future, there will be a considerable number of studies that, rather of focusing on eliminating beam squint, would instead focus on using it in a certain way in order to accomplish a certain goal. As a direct consequence of this, mmwave communication systems will become simpler and less expensive. Table 2 is a synopsis of the studies that were conducted on beam squint by a number of different people, and it lists the conclusions from that research. We have accomplished all that is within our ability to answer the most pertinent results and remarks, in addition to the several types of methods that were used in these studies. It was found that the tradeoff between the array size and the reduction of the impacts of beam squint was significantly similar and directly related to the cost and design complexities of the hardware that was employed for these systems. This was found to be the case after it was discovered that these two factors were directly related to each other.

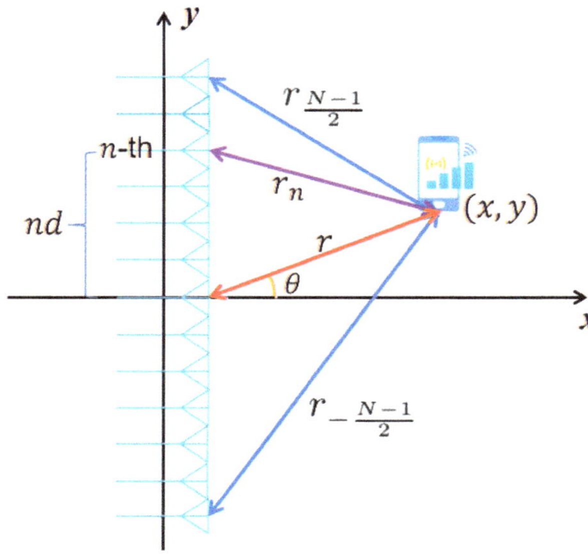

Figure 17. Near-field system model [67].

Table 2. Comparison among various studies conducted in the field of beam squint from 2012 to 2022.

Ref.	Year	Method	Results	Remarks
[46]	2012	TTD instead of phase shifters	TDD provided good result in overcoming beam squint	Increasing complexity and cost
[47]	2012	Negative Group Delay Circuits	Reduced the beam squint	Hardware complexity
[49]	2013	Three beamforming codebook design schemes	Reduced the beam squint	Increases the number of phase shifters, which increases power usage and complexity
[50]	2016	Algorithm that ensures minimum gain for all frequencies and angles	Reduced the beam squint	Greatly increases the size of codebook, which leads to high latency
[51]	2018	Alamuti-based beamforming scheme	May successfully fix the beam squint and increase throughput performance.	Proposed beam pattern optimization involves eigenvalue and -vector computation, the computation complexity of this scheme grows with the number of antennas
[52]	2019	Proposed subcarrier-to-beam allocation scheme	Exploited beam squint to enhance system performance	Hardware complexity since it is applied in digital domain
[53]	2019	Frequency-division duplex (FDD) massive MIMO-OFDM systems with hybrid analog/digital precoding using a channel estimation scheme	Provided more accurate channel estimation than other conventual methods	The assignment of pilot subcarriers and the design of training beams were not considered

Table 2. *Cont.*

Ref.	Year	Method	Results	Remarks
[54]	2019	Shift-invariant block-sparsity-based compression sensing (CS) algorithm	Provided more accurate channel estimation than other conventual methods	-
[55]	2020	Tansec waveform with nonlinear frequency modulation to account for carrier frequency variation	Beam direction error was reduced	Implemented in digital domain which increase complexity
[56]	2021	Algorithm for channel estimation using the greatest likelihood criterion in relation to the analog training beam and pilot subcarrier assignment	The proposed algorithm minimized the mean square error	-
[57]	2021	For RIS-aided wideband (mmwave) communications, phase shift design approaches are used to reduce the impact of beam squint in both LoS and NLoS scenarios	Beam squint in such systems was effectively reduced by the proposed phase shift architecture	-
[59]	2021	Virtual subarrays and TDD lines	Effective in removing beam squint	Increasing hardware complexity
[60]	2021	Advanced analog beamforming method proposed based on the Zadoff-Chu (ZC) sequence	The performance loss caused by the beam squint can be efficiently reduced using the suggested strategy	-
[61]	2021	Three-dimensional electromagnetic analysis software	Clarify how beam squint occurred in RF lens antenna and compare its effect with phased array antenna	-
[63]	2022	Transceiver design based on lens antenna subarray (LAS) and analog sub band filters	Results showed performance enhancement	Higher frequencies require high quality filters which increase hardware complexity
[65]	2020	Create multi-user communication plans that make use of squinting multi-beam properties and small multi-wideband antenna arrays	Effective method to exploit beam squint with multi-user systems	Squint-beam-based compact multi-wideband array for millimeter-wave communications
[67]	2022	Control the beam squint using time delays and using it for localization	Effective method to exploit beam squint in localization	-

5. Conclusions

In this paper, a literature review has been introduced to analyse the relative issues of the squint phenomena of mmwaves in antenna arrays for modern communication systems. It is found that such problems become very effective at high-frequency bands due to the dispersion effects of antenna arrays in terms of angle of arrival and gain variation, which directly effect the channel performance. For this purpose, the authors classified the main effective solutions to resolve such problems, which include beamforming, antenna geometry, and channel estimation algorithms. Another classification can be considered that exploits the beam squint to take advantages from squint phenomena to enhance the channel localization and capacity through splitting the beam to different directions. In spite of these classifications, beamforming, for example, based on digital processes, realizes an excellent solution to minimize the effect of beam squint on bit error rate and channel capacity, however; it consumes high complexity with excessive cost. The relative algorithm based on channel estimation could be considered the next revolutionary approach, whereby the system consumes less hardware complexity and cost. However, it is very limited when the channel environment becomes highly dynamic, which consumes high delay in the data processing. For this reason, it becomes unreliable with increasing traffic and signal attenuation. The third classification is assumed to be a solution method; however, it is not. It is indeed a suggestion to avoid the negative effects of such phenomena by splitting the beam of the antenna array to support channel localization and create multiuser communication channels. In view of the previous indications, a possible connection is found between the proposed classes to create a hybrid mixture to realize a high impact on the channel performance enhancement. This conclusion is built on the fact of dispersion, which is a relative problem with antenna arrays, channels, and subcarriers that cannot be resolved from one side and neglected from the other two. Therefore, for future research, a new technique must be recognized to find a complete solution that takes into the account beam squint effects at all communication system parts, thus leading us to fill the gaps in the developments in mmwave communication system for 5G and beyond.

Author Contributions: Investigation and original draft preparation, M.Q.A., supervision, methodology and review, A.H.M., Conceptualization, review and editing, R.A.A.-A. All authors have read and agreed to the published version of the manuscript.

Funding: Authors thank the support of the higher education ministry from Iraq.

Conflicts of Interest: There is no conflict of interest.

References

1. Cihat, S.; Muhammet, T.; Turgut, O. A Review of Millimeter Wave Communication for 5G. In Proceedings of the 2018 2nd International Symposium on Multidisciplinary Studies and Innovative Technologies (ISMSIT), Ankara, Turkey, 19–21 October 2018; IEEE: New York, NY, USA, 2018.
2. Ullah, H.; Nair, N.G.; Moore, A.; Nugent, C.; Muschamp, P.; Cuevas, M. 5G Communication: An Overview of Vehicle-to-Everything, Drones, and Healthcare Use-Cases. *IEEE Access* **2019**, *7*, 37251–37268. [CrossRef]
3. Duong, T.; Elkashlan, M. Millimeter-wave communications for 5G: Fundamentals. *IEEE Commun. Mag.* **2014**, *52*, 52–54.
4. Wei, L.; Hu, R.Q.; Qian, Y.; Wu, G. Key elements to enable millimeter wave communications for 5G wireless systems. *IEEE Wirel. Commun.* **2014**, *21*, 136–143.
5. van Berlo, B.; Elkelany, A.; Ozcelebi, T.; Meratnia, N. Millimeter Wave Sensing: A Review of Application Pipelines and Building Blocks. *IEEE Sensors J.* **2021**, *21*, 10332–10368. [CrossRef]
6. Mohanty, S.; Agarwal, A.; Agarwal, K.; Mali, S.; Misran, G. Role of Millimeter Wave for Future 5G Mobile Networks: Its Po-tential, Prospects and Challenges. In Proceedings of the 2021 1st Odisha International Conference on Electrical Power Engineering, Communication and Computing Technology (ODICON), Bhubaneswar, India, 8–9 January 2021; pp. 1–4.
7. Rappaport, T.S.; Sun, S.; Mayzus, R.; Zhao, H.; Azar, Y.; Wang, K.; Wong, G.N.; Schulz, J.K.; Samimi, M.; Gutierrez, F. Millimeter wave mobile communications for 5g cellular: It will work. *Access IEEE* **2013**, *1*, 335–349. [CrossRef]
8. Bogale, T.; Wang, X.; Le, L. Chapter 9–mm-Wave communication enabling techniques for 5G wireless systems: A link level perspective. *Mmwave Massive MIMO* **2017**, 195–225.
9. Wang, M.; Gao, F.; Jin, S.; Lin. An Overview of Enhanced Massive MIMO With Array Signal Processing Techniques. *IEEE J. Sel. Top. Signal Process.* **2019**, *13*, 886–901. [CrossRef]
10. Akyildiz, I.F.; Han, C.; Nie, S. Combating the Distance Problem in the Millimeter Wave and Terahertz Frequency Bands. *IEEE Commun. Mag.* **2018**, *56*, 102–108. [CrossRef]
11. Busari, S.A.; Huq, K.M.S.; Mumtaz, S.; Dai, L.; Rodriguez, J. Millimeter-Wave Massive MIMO Communication for Future Wireless Systems: A Survey. *IEEE Commun. Surv. Tutorials* **2017**, *20*, 836–869. [CrossRef]
12. Sarieddeen, H.; Saeed, N.; Al-Naffouri, T.Y.; Alouini, M.-S. Next Generation Terahertz Communications: A Rendezvous of Sensing, Imaging, and Localization. *IEEE Commun. Mag.* **2020**, *58*, 69–75. [CrossRef]
13. Al-Sadoon, M.; Patwary, M.; Parchin, N.; Aldelemy, A.; Abd-Alhameed, R. New Beamforming Ap-proach Using 60 GHz Antenna Arrays for Multi-Beams 5G Applications. *Electronics* **2022**, *11*, 1739. [CrossRef]
14. Xiao, M.; Mumtaz, S.; Huang, Y.; Dai, L.; Li, Y.; Matthaiou, M.; Karagiannidis, G.K.; Bjornson, E.; Yang, K.; Chih-Lin, I.; et al. Millimeter Wave Communications for Future Mobile Networks. *IEEE J. Sel. Areas Commun.* **2017**, *35*, 1909–1935. [CrossRef]
15. Du, J.; Valenzuela, R.A. How Much Spectrum is too Much in Millimeter Wave Wireless Access. *IEEE J. Sel. Areas Commun.* **2017**, *35*, 1444–1458. [CrossRef]
16. Andrews, J.G.; Buzzi, S.; Choi, W.; Hanly, S.V.; Lozano, A.; Soong, A.C.K.; Zhang, J.C. What will 5G be? *IEEE J. Sel. Areas Commun.* **2014**, *32*, 1065–1082. [CrossRef]
17. Andrews, J.G.; Bai, T.; Kulkarni, M.N.; Alkhateeb, A.; Gupta, A.K.; Heath, R.W. Modeling and Analyzing Millimeter Wave Cellular Systems. *IEEE Trans. Commun.* **2016**, *65*, 403–430. [CrossRef]
18. Swindlehurst, A.L.; Ayanoglu, E.; Heydari, P.; Capolino, F. Millimeter-wave massive MIMO: The next wireless revolution? *IEEE Commun. Mag.* **2014**, *52*, 56–62. [CrossRef]

19. Feng, W.; Li, Y.; Jin, D.; Su, L.; Chen, S. Millimeter-wave backhaul for 5G networks: Challenges and solutions. *Sensors* **2016**, *16*, 892. [CrossRef]
20. Li, J.; Niu, Y.; Wu, H.; Ai, B.; Chen, S.; Feng, Z.; Zhong, Z.; Wang, N. Mobility Support for Millimeter Wave Communications: Opportunities and Challenges. *IEEE Commun. Surv. Tutorials* **2022**, *24*, 1816–1842. [CrossRef]
21. Zugno, T.; Drago, M.; Giordani, M.; Polese, M.; Zorzi, M. Toward standardization of millimeter-wave vehicle-tovehicle networks: Open challenges and performance evaluation. *IEEE Communi. Mag.* **2020**, *58*, 79–85. [CrossRef]
22. Kutty, S.; Sen, D. Beamforming for Millimeter Wave Communications: An Inclusive Survey. *IEEE Commun. Surv. Tutorials* **2015**, *18*, 949–973. [CrossRef]
23. Bjornson, E.; Van der Perre, L.; Buzzi, S.; Larsson, E.G. Massive MIMO in sub-6 GHz and mm-Wave: Physical, practical, and use-case differences. *IEEE Wirel. Commun.* **2019**, *26*, 100–108. [CrossRef]
24. Mandloi, M.S.; Gupta, P.; Parmar, A.; Malviya, P.; Malviya, L. Beamforming MIMO Array Antenna for 5G-Millimeter-Wave Application. *Wirel. Pers. Commun.* **2022**, 1–20. [CrossRef]
25. Herd, J.S.; Conway, M.D. The Evolution to Modern Phased Array Architectures. *Proc. IEEE* **2015**, *104*, 519–529. [CrossRef]
26. Araújo, D.C.; Maksymyuk, T.; de Almeida, A.L.; Maciel, T.; Mota, J.C.; Jo, M. Massive MIMO: Survey and future research topics. *IET Commun.* **2016**, *10*, 1938–1946. [CrossRef]
27. Bogale, T.E.; Le, L.B.; Haghighat, A.; Vandendorpe, L. On the Number of RF Chains and Phase Shifters, and Scheduling Design with Hybrid Analog–Digital Beamforming. *IEEE Trans. Wirel. Commun.* **2010**, *15*, 3311–3326. [CrossRef]
28. Venkateswaran, M.V.; van der Veen, A. Analog Beamforming in MIMO Communications with Phase Shift Networks and Online Channel Estimation. *IEEE Trans. Signal Process.* **2010**, *58*, 4131–4143. [CrossRef]
29. Steyskal, H. Digital beamforming. In Proceedings of the 1988 18th European Microwave Conference, Stockholm, Sweden, 12–15 September 1988.
30. Heath, R.W.; González-Prelcic, N.; Rangan, S.; Roh, W.; Sayeed, A.M. An Overview of Signal Processing Techniques for Millimeter Wave MIMO Systems. *IEEE J. Sel. Top. Signal Process.* **2016**, *10*, 436–453. [CrossRef]
31. Dan; Dudgeon, E.; Johnson, D.H. *Array Signal Processing: Concepts and Techniques*; P T R Prentice Hall: Hoboken, NJ, USA, 1993.
32. Sohrabi, F.; Yu, W. Hybrid Analog and Digital Beamforming for mmWave OFDM Large-Scale Antenna Arrays. *IEEE J. Sel. Areas Commun.* **2017**, *35*, 1432–1443. [CrossRef]
33. Zhu, G.; Huang, K.; Lau, V.K.N.; Xia, B.; Li, X.; Zhang, S. Hybrid Beamforming via the Kronecker Decomposition for the Millimeter-Wave Massive MIMO Systems. *IEEE J. Sel. Areas Commun.* **2017**, *35*, 2097–2114. [CrossRef]
34. Gherekhloo, S.; Ardah, K.; Haardt, M. Hybrid Beamforming Design for Downlink MU-MIMO-OFDM Millimeter-Wave Systems. In Proceedings of the 2020 IEEE 11th Sensor Array and Multichannel Signal Processing Workshop (SAM), Hangzhou, China, 8–11 June 2020; pp. 1–5.
35. Khan, F.; Pi, Z.; Rajagopal, S. Millimeter-wave mobile broadband with large scale spatial processing for 5G mobile com-munication. In Proceedings of the 2012 50th Annual Allerton Conference on Communication, Control, and Computing (Allerton), Monticello, IL, USA, 1–5 October 2012; pp. 1517–1523.
36. Zhang, J.; Yu, X.; Letaief, K.B. Hybrid Beamforming for 5G and Beyond Millimeter-Wave Systems: A Holistic View. *IEEE Open J. Commun. Soc.* **2019**, *1*, 77–91. [CrossRef]
37. Cai, M.; Laneman, J.N.; Hochwald, B. Beamforming codebook compensation for beam squint with channel capacity con-straint. In Proceedings of the 2017 IEEE International Symposium on Information Theory (ISIT), Aachen, Germany, 25–30 June 2017; pp. 76–80.
38. Pan, X.; Li, C.; Yang, L. Wideband beamspace squint user grouping algorithm based on subarray collaboration. *EURASIP J. Adv. Signal Process.* **2022**, *2022*, 1–14. [CrossRef]
39. Available online: https://www.analog.com/en/analog-dialogue/articles/phased-array-antenna-patterns-part2.html (accessed on 1 June 2020).
40. Yao, J. Microwave Photonics. *J. Light. Technol.* **2009**, *27*, 314–335. [CrossRef]
41. Bonjour, R.; Singleton, M.; Leuchtmann, P.; Leuthold, J. Comparison of steering angle and bandwidth for various phased array antenna concepts. *Opt. Commun.* **2016**, *373*, 35–43. [CrossRef]
42. Cai, M. Modeling and Mitigating Beam Squint in Millimeter Wave Wireless Communication. Ph.D. Thesis, Graduate Program in Electrical Engineering, Notre Dame, Indiana, March 2012.
43. Rousseaux, O.; Leus, G.; Stoica, P.; Moonen, M. Gaussian maximum-likelihood channel estimation with short training se-quences. *IEEE Trans. Wirel. Commun.* **2005**, *4*, 2945–2955. [CrossRef]
44. Targonski, S.; Pozar, D. Minimization of beam squint in microstrip reflectarrays using an offset feed. In Proceedings of the IEEE Antennas and Propagation Society International Symposium. 1996 Digest, Baltimore, MD, USA, 21–26 July 1996; Volume 2, pp. 1326–1329.
45. Sharma, S.B.; Pujara, D.A.; Chakrabarty, S.B.; Singh, V.K. Removal of Beam Squinting Effects in a Circularly Polarized Offset Parabolic Reflector Antenna Using a Matched Feed. *Prog. Electromagn. Res. Lett.* **2009**, *7*, 105–114. [CrossRef]
46. Garakoui, S.K.; Klumperink, E.A.M.; Nauta, B.; van Vliet, F.E. Phased-array antenna beam squinting related to frequency dependency of delay circuits. In Proceedings of the 2011 41st European Microwave Conference, Manchester, UK, 10–13 October 2011; pp. 1304–1307.

47. Longbrake, M. True time-delay beam steering for radar. In Proceedings of the 2012 IEEE National Aerospace and Electronics Conference (NAECON), Dayton, OH, USA, 25–27 July 2012; pp. 246–249.
48. Alomar, W.; Mortazawi, A. Elimination of beam squint in uniformly excited serially fed antenna arrays using negative group delay circuits. In Proceedings of the 2012 IEEE International Symposium on Antennas and Propagation, Chicago, IL, USA, 8–14 July 2012; pp. 1–2.
49. Brilloun, L.; Sommerfeld, A. *Wave Propagation and Group Velocity*; Academic Press: Cambridge, MA, USA, 1960; pp. 113–137.
50. Liu, Z.; Rehman, W.U.; Xu, X.; Tao, X. Minimize Beam Squint Solutions for 60GHz Millimeter-Wave Communication System. In Proceedings of the 2013 IEEE 78th Vehicular Technology Conference (VTC Fall), Las Vegas, NV, USA, 2–5 September 2013; pp. 1–5.
51. Cai, M.; Gao, K.; Nie, D.; Hochwald, B.; Laneman, J.N.; Huang, H.; Liu, K. Effect of Wideband Beam Squint on Codebook Design in Phased-Array Wireless Systems. In Proceedings of the 2016 IEEE Global Communications Conference (GLOBECOM), Washington, DC, USA, 4–8 December 2016; pp. 1–6.
52. Liu, X.; Qiao, D. Space-Time Block Coding-Based Beamforming for Beam Squint Compensation. *IEEE Wirel. Commun. Lett.* **2018**, *8*, 241–244. [CrossRef]
53. Laurinavicius, I.; Zhu, H.; Wang, J.; Pan, Y. Beam Squint Exploitation for Linear Phased Arrays in a mmWave Multi-Carrier System. In Proceedings of the 2019 IEEE Global Communications Conference (GLOBECOM), Waikoloa, HI, USA, 9–13 December 2019; pp. 1–6.
54. Wang, B.; Jian, M.; Gao, F.; Li, G.Y.; Lin, H. Beam Squint and Channel Estimation for Wideband mmWave Massive MIMO-OFDM Systems. *IEEE Trans. Signal Process.* **2019**, *67*, 5893–5908. [CrossRef]
55. Wang, M.; Gao, F.; Gu, Y.; Flanagan, M.F. A Block Sparsity Based Channel Estimation Technique for mmWave Massive MIMO with Beam Squint Effect. In Proceedings of the ICC 2019-2019 IEEE International Conference on Communications (ICC), Shanghai, China, 20–24 May 2019; pp. 1–6.
56. Longman, O.; Solodky, G.; Bilik, I. Beam Squint Correction for Phased Array Antennas Using the Tansec Waveform. In Proceedings of the 2020 IEEE International Radar Conference (RADAR), Washington, DC, USA, 28–30 April 2020; pp. 489–493.
57. Noh, S.; Lee, J.; Yu, H.; Song, J. Design of Channel Estimation for Hybrid Beamforming Millimeter-Wave Systems in the Presence of Beam Squint. *IEEE Syst. J.* **2021**, *16*, 2834–2843. [CrossRef]
58. Chen, Y.; Chen, D.; Jiang, T. Beam-Squint Mitigating in Reconfigurable Intelligent Surface Aided Wideband MmWave Communications. In Proceedings of the 2021 IEEE Wireless Communications and Networking Conference (WCNC), Nanjing, China, 29 March–1 April 2021; pp. 1–6.
59. Liaskos, C.; Nie, S.; Tsioliaridou, A.; Pitsillides, A.; Ioannidis, S.; Akyildiz, I. A New Wireless Communication Paradigm through Software-Controlled Metasurfaces. *IEEE Commun. Mag.* **2018**, *56*, 162–169. [CrossRef]
60. Gao, F.; Wang, B.; Xing, C.; An, J.; Li, G.Y. Wideband Beamforming for Hybrid Massive MIMO Terahertz Communications. *IEEE J. Sel. Areas Commun.* **2021**, *39*, 1725–1740. [CrossRef]
61. Wang, Z.; Liu, J.; Wang, J.; Yue, G. Beam squint effect on high-throughput millimeter-wave communication with an ul-tra-massive phased array. *Front. Inform. Technol. Electron. Eng.* **2021**, *22*, 560–570. [CrossRef]
62. Park, S.-H.; Kim, B.; Kim, D.K.; Dai, L.; Wong, K.-K.; Chae, C.-B. Beam Squint in Ultra-wideband mm-Wave Systems: RF Lens Array vs. Phase-Shifter-Based Array. *arxiv* **2021**, arXiv:2112.04188.
63. Lorentz, H.A. *The Theory of Electrons: And Its Applications to the Phenomena of Light and Radiant Heat*; Wentworth Press: Sydney, Australia, 2019.
64. Afeef, L.; Kihero, A.B.; Arslan, H. Novel Transceiver Design in Wideband Massive MIMO for Beam Squint Minimization. *arxiv* **2022**, arXiv:2207.04679.
65. Lin, Z.; Niu, H.; An, K.; Wang, Y.; Zheng, G.; Chatzinotas, S.; Hu, Y. Refracting RIS-Aided Hybrid Satellite-Terrestrial Relay Networks: Joint Beamforming Design and Optimization. *IEEE Trans. Aerosp. Electron. Syst.* **2022**, *58*, 3717–3724. [CrossRef]
66. Pan, X.; Li, C.; Hua, M.; Yan, W.; Yang, L. Compact Multi-Wideband Array for Millimeter-Wave Communications Using Squint Beams. *IEEE Access* **2020**, *8*, 183146–183164. [CrossRef]
67. Lin, Z.; Lin, M.; Wang, J.B.; de Cola, T.; Wang, J. Joint Beamforming and Power Allocation for Satellite-Terrestrial Integrated Networks with Non-Orthogonal Multiple Access. *IEEE J. Sel. Top. Signal Process.* **2019**, *3*, 657–670. [CrossRef]
68. Luo, H.; Gao, F. Beam Squint Assisted User Localization in Near-Field Communications Systems. *Arvix* **2022**, arXiv:2205.11392.

electronics

MDPI

Article

Ultra-Broadband Angular-Stable Reflective Linear to Cross Polarization Converter

Bianmei Zhang [1,2], Chenghui Zhu [1], Ran Zhang [1], Xiaofan Yang [3], Ye Wang [1] and Xiaoming Liu [1,4,*]

1 School of Physics and Electronic Information, Anhui Normal University, Wuhu 241002, China
2 School of Electrical and Electronic Engineering, Anhui Institute of Information Technology, Wuhu 241002, China
3 The State Key Laboratory of Complex Electromagnetic Environment Effects on Electronic and Information System, Luoyang 471004, China
4 Wuhu CEPREI Information Industry Technology Research Institute, Wuhu 241002, China
* Correspondence: xiaoming.liu@ahnu.edu.cn

Abstract: An ultra-broadband angular-stable reflective linear to cross polarization converter based on metasurface is developed. The unit cell of the converter is formed by a slant end-loaded H-shaped resonator. The slant arrangement is to create polarization conversion. The end-loaded stubs are useful for miniaturization and the slots are responsible for enlarging bandwidth. The simulated results show that the polarization conversion ratio of the proposed design is better than 90% in the range of 9.83–29.37 GHz, corresponding to a relative bandwidth of 99.69%. It is also demonstrated that the mean polarization conversion ratio is larger than 80% even though the incident angle reaches 40° for both x-polarized and y-polarized incidences. To validate the design, a prototype of the proposed structure is fabricated and measured. Satisfactory agreement has been observed between measurement and simulation. Compared with the designs in the literature, the developed converter exhibits good performance of high efficiency, ultra-broadband and angular stability. Potential applications can be expected in polarization-controlled devices, stealth surfaces, antennas, etc.

Keywords: polarization converter; ultra-broadband; angular stability; metasurface; polarization conversion ratio

Citation: Zhang, B.; Zhu, C.; Zhang, R.; Yang, X.; Wang, Y.; Liu, X. Ultra-Broadband Angular-Stable Reflective Linear to Cross Polarization Converter. *Electronics* **2022**, *11*, 3487. https://doi.org/10.3390/electronics11213487

Academic Editors: Naser Ojaroudi Parchin, Mohammad Ojaroudi and Raed A. Abd-Alhameed

Received: 8 October 2022
Accepted: 25 October 2022
Published: 27 October 2022

1. Introduction

The polarization of a train of electromagnetic (EM) waves refers to the oscillating direction of its electric field in the plane perpendicular to the propagation direction [1]. Effective controlling or manipulating of polarization state is frequently required in many applications [2,3]. Traditional methods for polarization control include making uses of the birefringence effect and optical activity of natural materials [4–7]. However, these methods usually lead to a bulky thickness. Over the past decade, great efforts were devoted to the investigation of planar periodical structures, also called metasurfaces (MSs) due to their fascinating functionalities that cannot be found in conventional materials, such as negative refraction [8], superlens [9], invisibility cloak [10,11], perfect absorption [12] and so on. Metasurfaces also provide an efficient approach to polarization control in sub-wavelength scale [13–15]. In particular, by utilizing this method, the thickness may be significantly reduced.

In recent years, polarization converters based on anisotropic MSs have been intensively studied. In general, MS-based polarization converters can be classified into two types, i.e., the transmission type and the reflection type. Many of them have been reported over the microwave [16–18] and terahertz [19,20] ranges. For the transmission type, broadband polarization converters are frequently realized by staking multilayer structures [21–23]. However, its complex structure makes it less convenient to be fabricated. In contrast, the reflection type is capable of creating high-efficiency and broadband polarization conversion using monolayer MSs [24–29].

For instance, Zheng et al. [26] developed a polarization converter based on a modified H-shaped resonator that achieved linear polarization conversion with a bandwidth <61%. Li et al. [27] proposed another reflective polarization converter based on the square splitring resonators. The bandwidth was increased to 94% for polarization conversion ratio (PCR) greater than 80%. However, it is difficult for this design to maintain high conversion efficiency over a wide bandwidth. To increase the conversion efficiency, a perforated array was used [28,29], and the bandwidth could be as broad as 111%. Meanwhile, the PCR was better than 89% over the whole range. To further broaden the bandwidth and produce better PCR, Jia et al. [30] presented two cascaded dielectric-layer structure. Unfortunately, it only demonstrated a good efficiency for normal incidence.

Improvements have also been made to broaden both the bandwidth and the angular stability [31–36]. Karamirad et al. [31] introduced an oval pattern periodic array based on multiple plasmon resonances and surface magnetic field distributions. The bandwidth of this design was about 67% with an angular stability of 30°. Another design [32] using cross-shaped resonators enhanced the angular stability to 40°, but the bandwidth was not sufficiently wide. Wu et al. [36] presented a broadband design that consisted of hollow oval sheet and rectangle patch, providing a bandwidth up to 90% for the first band. However, this design was very sensitive to the incident angle, and only operated efficiently with 25° angular stability.

It is seen that, for a polarization converter, there is a dilemma between bandwidth, efficiency and angular stability. Therefore, investigating the development of easy fabrication, wide bandwidth and large angle stability cross-polarization converter is still worthy of a great deal of effort.

In this work, an ultra-broadband angular stable linear polarization-converter based on MS is developed. It is composed of a single layer slant end-loaded H-shaped resonator. The slant arrangement is to create polarization conversion. The end-loaded stubs are useful for miniaturization. Moreover, the slots are responsible for enlarging bandwidth. It will be demonstrated that the PCR efficiency of this design is better than 90%. Even when the incident angle reaches 40°, the mean PCR remains above 80%. It will also be demonstrated that a good tradeoff between bandwidth, PCR and angular stability has been obtained. The fabricated prototype confirms a good agreement between simulation and measurement.

2. Design and Simulation

The unit cell evolves from a simple slant dipole, as shown in Figure 1a. Following on, two U-shaped stubs are loaded with each end of the dipole, so that the unit cell is partly miniaturized without destroying polarization conversion, see Figure 1b. To enlarge its bandwidth, two slots are cut at the ends, forming an H-shaped resonator, as depicted in Figure 1c, cutting slots in a common way to enhance bandwidth in antenna design [37]. This technique is used in this design to enhance the bandwidth of polarization conversion. The dielectric layer is PTFE having a relative dielectric constant (ε_r) of 2.2 and a loss tangent ($\tan\delta$) of 0.0009. The thickness (h) of the substrate is 2.5 mm. The metallic layers are copper with thickness of 0.035 mm and electrical conductivity of $\sigma = 5.8 \times 10^7$ S/m. By using low loss substrate, it is possible to create a higher conversion efficiency. The resultant dimensions of the structure are p = 6 mm, s = 6.28 mm, w = 0.28 mm, g = 2.2 mm, l_1 = 1.61 mm, l_2 = 1.62 mm, l_3 = 0.65 mm. The unit cell has an overall size of $6 \times 6 \times 2.5$ mm^3.

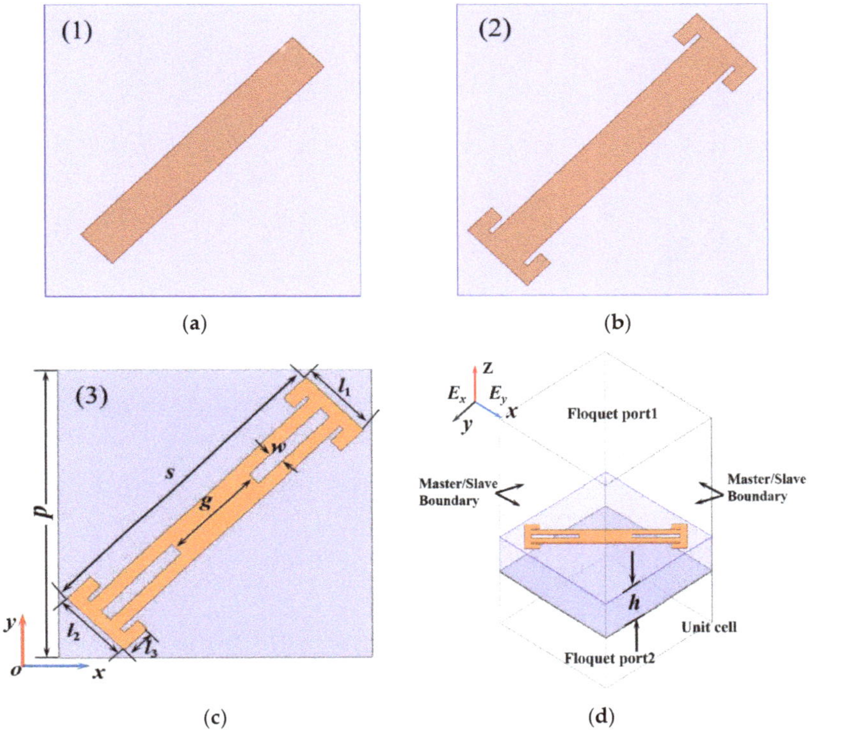

Figure 1. Evolution of the unit cell. (**a**) The slant dipole; (**b**) the end-loaded dipole; (**c**) the front view of the proposed polarization converter unit cell; (**d**) the schematic diagram of the port setting and boundary conditions for simulation.

A series of numerical simulations by using Ansoft HFSS are conducted to check the design. In the simulation model, the four boundaries are assigned as master/slave periodic conditions, as shown in Figure 1d. Since the unit cell shows mirror symmetry along its diagonal line, only the amplitudes of r_{xx} and r_{yx} along with their phase difference $\Delta\varphi_{xy} = \arg(r_{xx}) - \arg(r_{yx})$ under normal incidence for x-polarization are plotted in Figure 2. It is seen that the amplitudes of r_{xx} are smaller than -10 dB and that of r_{yx} are better than -1 dB in the range of 9.90–29.34 GHz. The phase differences $\Delta\varphi_{xy}$ are plotted in Figure 2b, where it is seen that $\Delta\varphi_{xy}$ is close to $\pm(2n+1)90^\circ$ (n is an integer) in the whole frequency region. These simulation results indicate that x-polarization at normal incidence is converted into y-polarization within the operating band.

Another parameter used to characterize polarization conversion is the PCR of the reflected wave, which is usually defined as [26]:

$$\mathrm{PCR} = r_{yx}^2/(r_{yx}^2 + r_{xx}^2) = 1 - r_{xx}^2/(r_{yx}^2 + r_{xx}^2). \quad (1)$$

The calculated PCR of the proposed design is plotted in Figure 3a. It is observed that the PCR is better than 90% in the range of 9.83–29.37 GHz, and reaches nearly 100% at the three resonance frequencies. The relative bandwidth is up to 99.69%, being broader than the designs in the literature [31–36]. Additionally, Figure 3b shows the polarization azimuth angle α to describe the angle between the major polarization axis and x-axis, which can be calculated as [25]:

$$\alpha = \arctan(r_{yx}/r_{xx}) \quad (2)$$

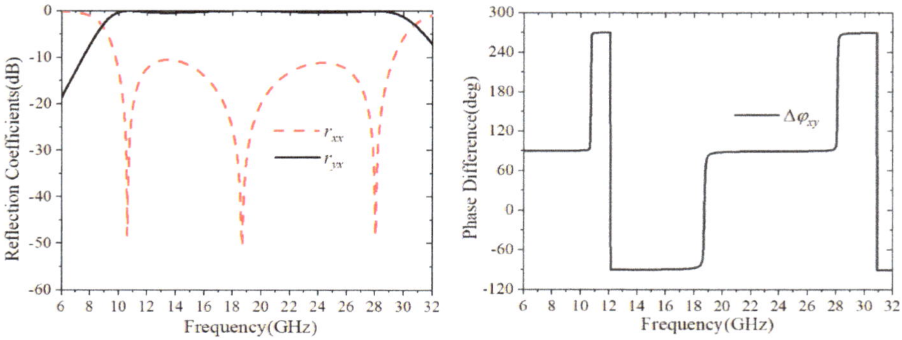

(**a**) (**b**)

Figure 2. The simulated results of the proposed polarization converter under normal incidence for x-polarization. (**a**) The magnitudes of r_{xx} and r_{yx}; (**b**) the phase difference $\Delta\varphi_{xy}$ between r_{xx} and r_{yx}.

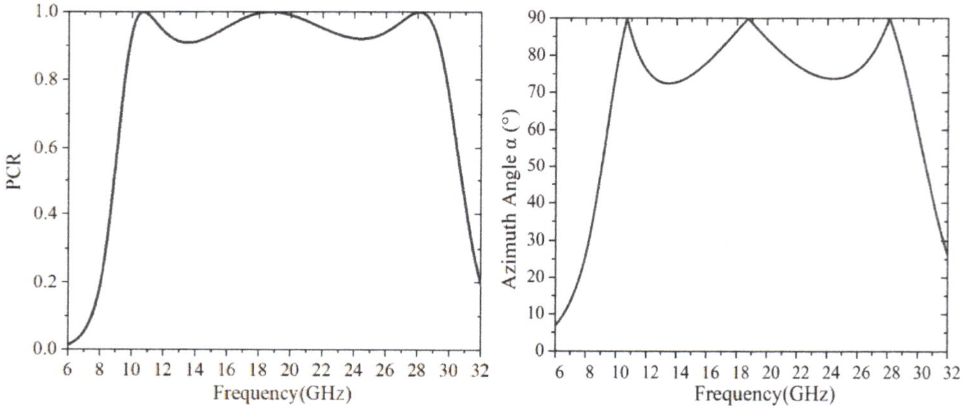

(**a**) (**b**)

Figure 3. The simulated results for (**a**) PCR; (**b**) polarization azimuth angle α.

This is a more intuitive parameter describing frequency dependent polarization conversion. It is seen from Figure 3b that the polarization azimuth angle α is larger than 72° in the range of 9.83–29.37 GHz. The values of α approach 90° at three resonance frequencies of 10.68, 18.74 and 28.10 GHz. These results further confirm that the high-efficiency cross polarization conversion has been realized from x-polarized to y-polarized over the broad frequency bands [27].

A comparison of PCRs for each cell in Figure 1 is plotted in Figure 4. It can be found that the slant dipole array is capable of stimulating polarization rotation. However, the bandwidth of PCR does not appear to be any better compared with the other two design, though the structure is simpler. The end is then loaded two U-shaped stubs to minimize the design, which leads to an increase in resonance points. The resultant bandwidth of this structure is about 94.2% with PCR as high as 0.9. It is recognized that a wider bandwidth may be obtained by introducing slots to the structure [33]; two slots are cut in the slant dipole, as shown in Figure 1c. Such modification increases the bandwidth to 99.69%.

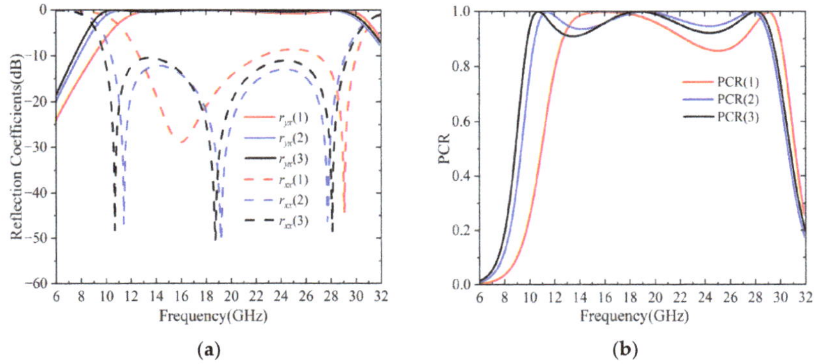

Figure 4. Simulation results for unit cells. (**a**) Reflective coefficients; (**b**) PCRs.

Another merit of the polarization converter is its angular stability. The PCR is assumed to be dependent on the incident angle. As shown in Figure 5, the PCRs for different incident angles are plotted for x-polarized and y-polarized incidences. It is apparent that the PCR bandwidth of the polarization converter gradually reduces when the incident angle increases from 0° to 40°. In the higher frequency region (around 26 GHz), the influence due to the increase in the incident angle is more pronounced, which may be attributed to the destructive interference caused by the extra path traveled by the wave inside the substrate [38]. Even though there is a decrease in the PCR bandwidth, it is seen from Figure 5a that a wide angular stability up to 20° can be observed from 10.4 to 26 GHz for the x-polarized wave. This frequency range is defined in terms of 90% PCR. For the y-polarized wave, the corresponding range is from 10.33 to 25.93 GHz. Therefore, it is seen that the polarization converter is independent from the polarization state of the incident wave. Moreover, even for 30° incidence, the PCR can be more than 85% from the range of (10.29–24.66 GHz). Furthermore, the mean PCR is still better than 80%, even though the incident angle is up to 40°. This is a piece of good evidence showing that the proposed converter has a satisfactory angular stability over the range of 0–40°. It should be noted that the proposed polarization converter has a cell periodicity of $0.197\lambda_{max}$ and a thickness of $0.082\lambda_{max}$, where λ_{max} is the free-space wavelength corresponding to the lowest frequency in the frequency range of 9.83–29.37 GHz.

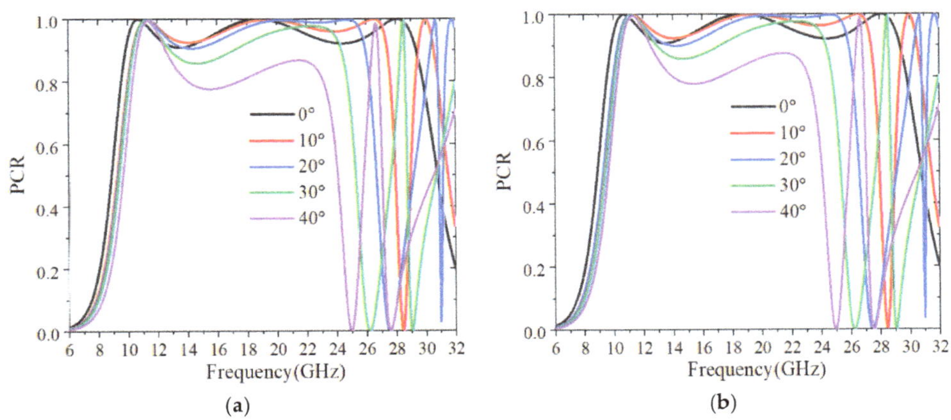

Figure 5. The PCR under different oblique incidences for (**a**) x-polarized wave; (**b**) y-polarized wave.

Actually, this design loads a U-shaped stub at each end of the dipole as resonator so as to reduce side length, which is a typical technique for miniaturization. It is recognized that the resonance characteristic of the structure is dependent on the geometric parameters of structure [31,38]. By decreasing the size of the unit cell, the resonance characteristic is better met. To study the influence of the size of the unit cell on the incident angle stability, the amplitudes of r_{xx} for different parametric sweep on incident angle vs. the size of unit cell are plotted in Figure 6. It can be seen that with the decrease in the size of unit cell, the resonant frequency of r_{xx} at lower frequencies almost remains stable, while that of r_{xx} at higher frequencies shifts towards higher frequencies, when the incident angle is kept constant. Moreover, the dip values of the resonance frequencies are better at different incident angles when the size of the unit cell gradually decreases, especially for middle resonance frequencies. It is evident that this structure will produce better PCR or wider bandwidth for the miniaturization of the unit cell, as shown in Figure 7. Therefore, the wide incident angle stability of the proposed converter can be attributed to the miniaturization of the unit cell.

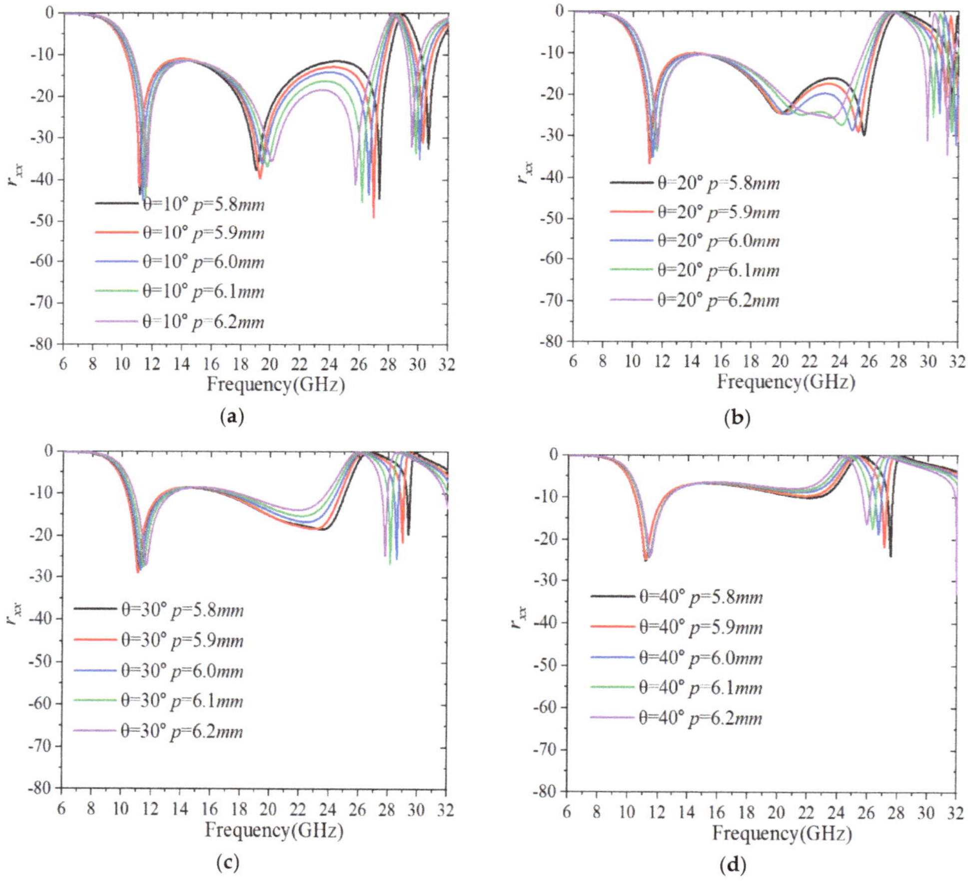

Figure 6. The amplitudes of r_{xx} for normal x-polarization incidence. (**a**) θ = 10°; (**b**) θ = 20°; (**c**) θ = 30°; (**d**) θ = 40°.

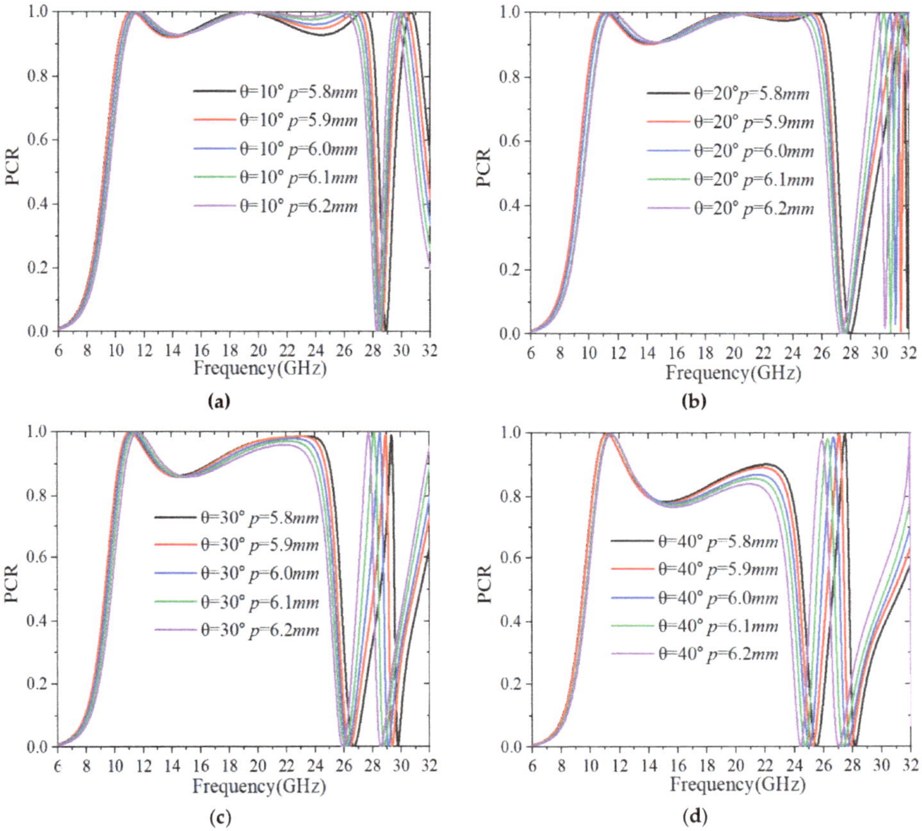

Figure 7. The PCRs for normal x-polarization incidence. (**a**) θ = 10°; (**b**) θ = 20°; (**c**) θ = 30°; (**d**) θ = 40°.

3. Analysis and Discussion

3.1. Theoretical Analysis of Linear to Cross Polarization

To understand the underlying physical mechanisms of the proposed polarization converter, a new coordinate system (the u-axis and v-axis) is usually defined to analyze the structural anisotropy. They are obtained via rotating the x-axis and y-axis counterclockwise by 45°. The x-polarized incident EM wave ($\vec{E}^{i}_{x}$) can be decomposed into two components along the u-axis ($\vec{E}^{i}_{u}$) and v-axis ($\vec{E}^{i}_{v}$), as shown in Figure 8a. Thus, the incident and reflected electric fields can be expressed as [26]:

$$\vec{E}^{i}_{x} = \vec{x}\left|\vec{E}^{i}_{x}\right| = \vec{x}E_0e^{jkz} = \vec{E}^{i}_{u} + \vec{E}^{i}_{v} = \vec{u}\left|\vec{E}^{i}_{u}\right| + \vec{v}\left|\vec{E}^{i}_{v}\right|. \tag{3}$$

The reflection coefficients aligned along the u-direction and v-direction can be defined as r_{uu} and r_{vv}, respectively. Then, the reflected field can be written as [31]:

$$\vec{E}^{r}_{x} = \vec{E}^{r}_{u} + \vec{E}^{r}_{v} = \vec{u}r_{uu}\left|\vec{E}^{i}_{u}\right|e^{j(kz+\varphi_{uu})} + \vec{v}r_{vv}\left|\vec{E}^{i}_{v}\right|e^{j(kz+\varphi_{vv})}. \tag{4}$$

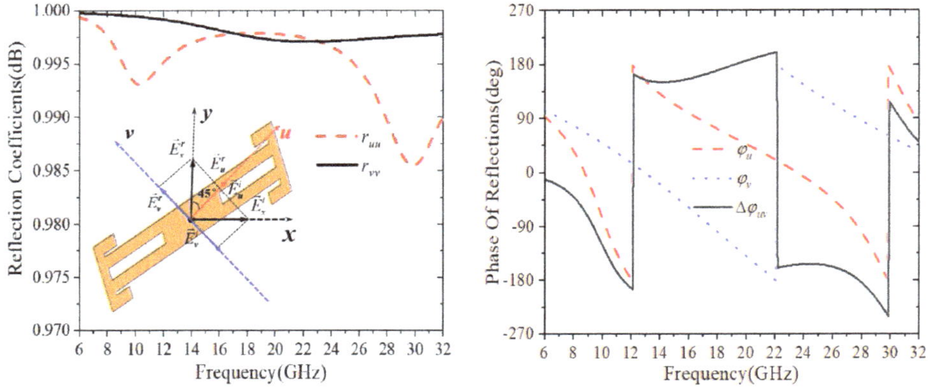

(a) (b)

Figure 8. Simulated reflection coefficients of the proposed polarization converter under u- and v-polarized normal incidences. (**a**) Amplitude; (**b**) Phase.

For the lossless case, both of the incident and reflected waves can be regarded as a composite wave composed of u- and v-polarized components with equal amplitude for x-polarized incidence. If the reflection coefficients and phase difference between r_{uu} and r_{vv} satisfy the conditions of [33]:

$$\begin{cases} r_{uu} = r_{vv} = r \\ \Delta\varphi_{uv} = \varphi_{uu} - \varphi_{vv} = \pi \end{cases} . \tag{5}$$

a perfect cross-polarization conversion will be formed. If $\Delta\varphi_{uv} \neq 0°$ or $\Delta\varphi_{uv} \neq \pm 180°$, the reflected wave will be elliptically polarized. Specifically, when the phase difference $\Delta\varphi_{uv} = 90°$, the reflected wave will be a circular polarization one, which means that linear to circular polarization conversion is obtained.

Illuminating the structure with polarization along the u- and v-axes, respectively, the reflection coefficients (r_{uu} and r_{vv}) can be obtained, as plotted in Figure 8a. It is seen that the magnitudes of r_{uu} and r_{vv} are both nearly equal to one (better than 0.98) over the simulation frequency range (6–32 GHz). In addition, there are three minimum values located at 10.36, 22.12 and 29.92 GHz, indicating that three resonance modes are excited by u-polarized and v-polarized cases, respectively. In fact, the cell of the polarization converter can be equivalent to an LC parallel resonator, wherein the equivalent capacitance C are caused by the charge accumulations in the gaps between the adjacent metal patches, and the inductances are created by the metal patches of the polarization converter [36]. Due to the anisotropy of the structure, the values of the equivalent capacitance C and inductance L are both different at different resonant frequencies, which implies that the Q values of these resonant modes are also different [24,36]. However, the phase delay along the u-direction and v-direction has to be $\pm 180°$ (n is an integer) to form a cross polarization converter.

The phase variations and phase difference of r_{uu} and r_{vv} are plotted in Figure 8b. It is seen that there are different phase variations for r_{uu} and r_{vv} at three resonance modes. Evidently, φ_u and φ_v decrease rapidly at the first resonant mode (10.36 GHz), resulting in large phase differences, which make the phase difference $\Delta\varphi_{uv}$ fluctuate around $\pm 180°$ before the appearance of the next resonance mode. Subsequently, the phase difference $\Delta\varphi_{uv}$ can be kept close to $\pm 180°$ for the second resonant mode (22.12 GHz) due to a smaller Q value. However, φ_u and φ_v quick asynchronously for the third resonant mode (29.92 GHz), which broke the relative stability of $\Delta\varphi_{uv}$. Therefore, the phase difference $\Delta\varphi_{uv}$ is close to $\pm 180°$ in the frequency range of 10.36–29.92 GHz, showing that the anticipated cross polarization conversion has been realized in this ultra-wide frequency range.

To elaborate further on the physical mechanism behind the polarization conversion, the surface current distributions on the top layer and metallic ground plane are investigated for normal incidence in response to the *v*- and *u*-components with various resonant frequencies of 10.36, 22.12 and 29.92 GHz, as shown in Figure 9. It is seen that the current directions on the top and bottom metallic layer are anti-parallel at 10.36 and 22.12 GHz, which implies that a strong magnetic field is generated inside the substrate sandwiched between the two layers. Moreover, a current loop is formed in the intermediate dielectric layer, which is known as magnetic resonance. In contrast, the surface currents of top layer are parallel to those on the background sheet at 29.92 GHz, corresponding to electric resonance. Actually, the ultra-broadband enhancement operation results from the superstition of multiple PCR peaks around resonance frequencies. Therefore, multiple resonances are vital to realize high-efficiency and ultra-broadband cross-polarization conversion.

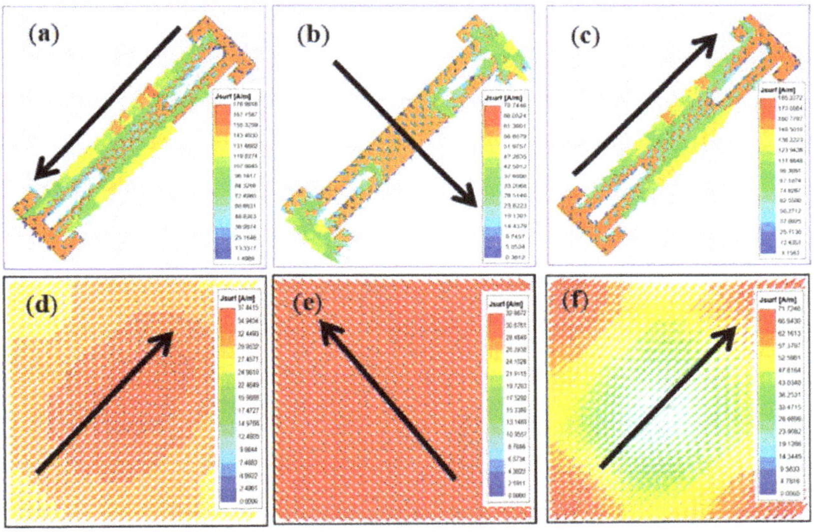

Figure 9. Surface current distributions on top layer and metallic ground plane for normal incidence in response to the *v*- and *u*-components with various resonant frequencies of (**a**,**b**) 10.36 GHz; (**c**,**d**) 22.12 GHz; (**e**,**f**) 29.92 GHz.

3.2. *Parametric Analysis*

3.2.1. Dependence of PCR on Unit Cell Periodicity

A systematic parametric study has been conducted to analyze the sensitivity of some critical dimensions. Since the periodicity p and substrate thickness h of unit cell have a larger dimension versus the length g, s, l_1, l_2 and l_3 or width w of unit cell, the step is set to 100 μm for p and h, but the length g is less sensitive for dimensional changes, the step is also set to 100 μm. A much smaller step (20 μm) is applied for other parameters. Following this, to investigate the relationship between the sensitivity of each parameter and fabrication accuracy, the PCRs are plotted for various parameters in Figure 10. The periodicity p of unit cell is varied in the range of 5.9–6.1 mm in a step of 0.1 mm, while other parameters are kept constant, as shown in Figure 10a. It is evident that the bandwidth of the working band is decreasing with the increase from 5.9 to 6.1 mm, while the polarization conversion efficiency increases slightly. Considering both operating bandwidth and the performance, the final value of p is chosen as 6 mm.

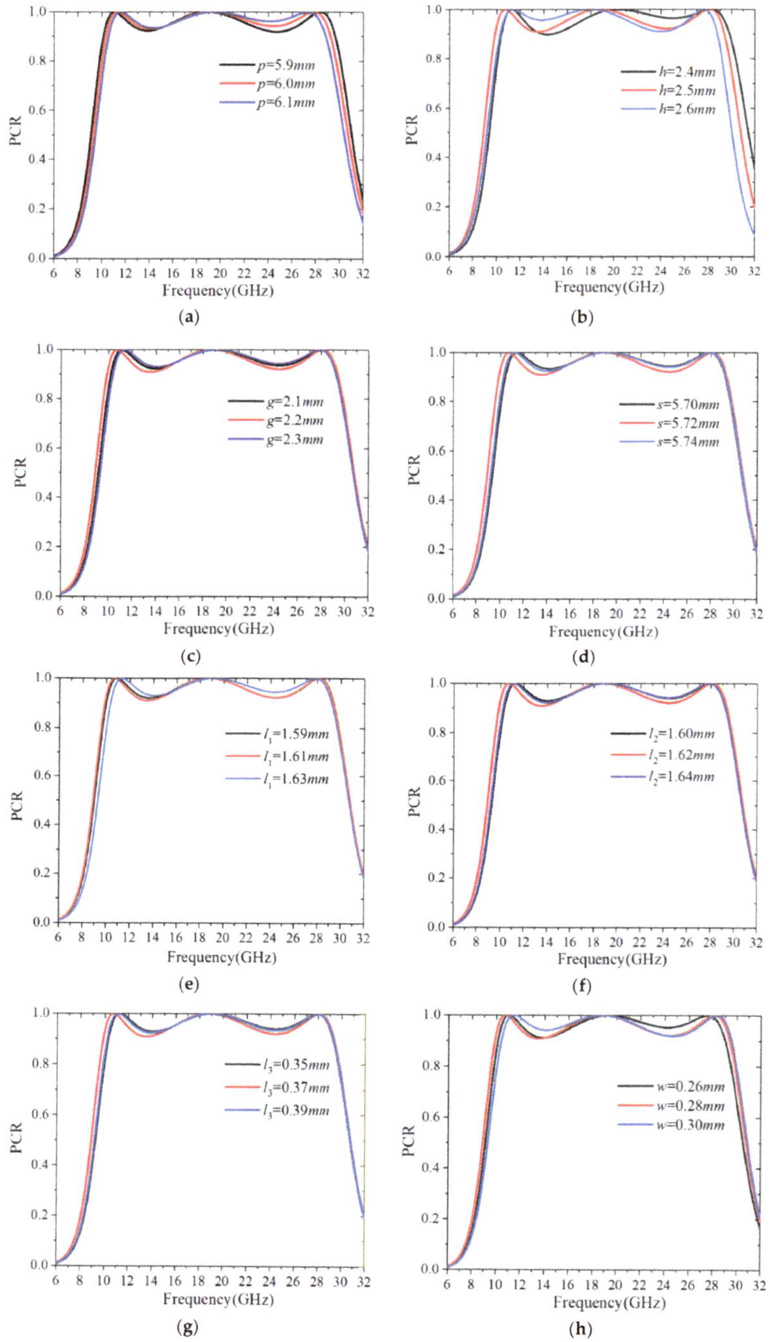

Figure 10. The PCR variation dependence on different parameters of the unit cell. (**a**) p; (**b**) h; (**c**) g; (**d**) s; (**e**) l_1; (**f**) l_2; (**g**) l_3; (**h**) w.

3.2.2. Dependence of PCR on Substrate Thickness

In addition, in order to make the working frequency band meet the requirements of specific applications, the changes of dielectric layer thickness h and other parameters of the unit cell are simulated. Figure 10b shows the PCR of the proposed converter as a function of h in the range of 2.4–2.6 mm without altering the other parameters. Obviously, by increasing h, the operating frequency of the proposed converter is shifted to a lower frequency range. Accordingly, the polarization conversion becomes strong reaching to almost 1 at low frequencies when h increases from 2.4 to 2.6 mm. Contrary to PCR in low frequencies, the PCR reduces to 0.9 with increasing h in high frequencies. Considering the desired broadband performance and commercially available thicknesses of the substrate layer, h = 2.5 mm is concluded as the optimized dimension.

3.2.3. Dependence of PCR on Unit Cell Length

After the p and h are determined, Figure 10c–g gives the PCR for various parameters versus the length g, s, l_1, l_2 and l_3 of unit cell. When g varies from 2.1 to 2.3 mm in 0.1 mm steps, and s, l_1, l_2 and l_3 increase in 0.02 mm steps, it is evident that in the 14–22 GHz frequency range, the polarization conversion almost remains stable. However, the data curves of the PCR still have a slight shift within the whole frequencies. Furthermore, the response of the PCR for different values of width w of unit cell is also depicted in Figure 10h. When w changes in the range of 0.26–0.3 mm, the polarization conversion efficiency increases slightly at low band, and plays an opposite role at high band. These simulation results imply that 10 μm fabrication accuracy is sufficiently good for this design. The design parameters are determined while considering both operating bandwidth and performance. Based on the above analysis, we can obtain the optimized structure shown in Figure 1.

4. Experimental Results

To verify the design, a prototype is fabricated using the standard print circuit board (PCB) technique. It consists of 41 × 41 unit cells over an area of 246 mm × 246 mm. The fabricated sample is examined by an industrial microscope, as shown in Figure 11. It can be observed that the fabrication accuracy is well within 10 μm, which is sufficient to obtain the stability of bandwidth and angular incidence.

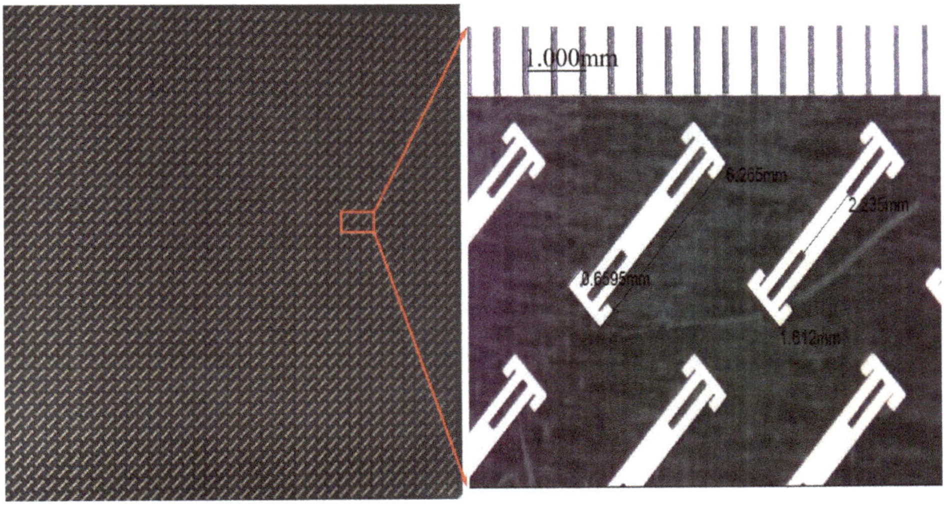

Figure 11. The fabricated sample and its unit cells under an industrial microscope.

The measurement setup is shown in Figure 12. Two horn antennas are connected to a vector network analyzer (Ceyear AV3672D). Radar absorbing materials (RAM) are mounted around the structure to avoid diffraction and spillover from the edge. The horn antennas are placed at the same height to ensure that the EM wave can be illuminated to the center of the structure. For the reflection coefficients measurement, the separation angle between two antennas is set at 5°, which corresponds to the normal incidence measurement in the experiment. By rotating the receiving horn antenna, the test setup is capable of dealing with both horizontal and vertical polarization, so the r_{xx} and r_{yx} can be measured. Moreover, in order to obtain the background, a sheet metal of the same size as the sample is measured to calibrate the system.

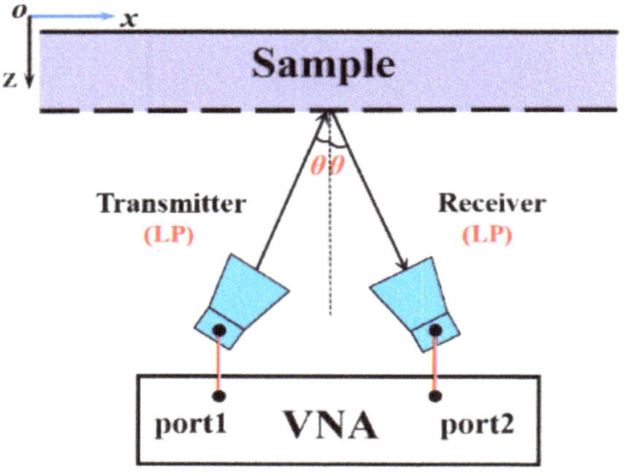

Figure 12. Structure diagram of the measurement setup.

Due to the limitation of our measurement components, the measurement on reflection coefficients is conducted from 15 GHz to 30 GHz. The measured co-polarized and cross-polarized reflection coefficients and the corresponding calculated PCRs at a different oblique incident angle are plotted in Figure 13. It can be seen from Figure 13 that the experimental results are in good agreement with the simulated ones. It has to be mentioned that normal incidence is not measured because of the blockage between the transmitting and receiving horns. Instead, the incident angle of 5° is measured trying to mimic normal incidence. Obviously, at near-normal incidence (oblique angle of 5°), r_{xx} is smaller than -10 dB and r_{yx} is higher than -3 dB in the frequency range of 15–27.42 GHz. Therefore, the calculated PCR is higher than 90%. These results solidly confirm that the illuminated x-polarized wave has been efficiently converted to y-polarized wave after reflection from the converter. Although the PCR decreases with the increase in incident angle, it is always larger than 78% in frequency ranges of 15–23.21 GHz even for a 40° incident angle. Such a result indicates that the proposed converter is insensitive for a wide range of incident angles.

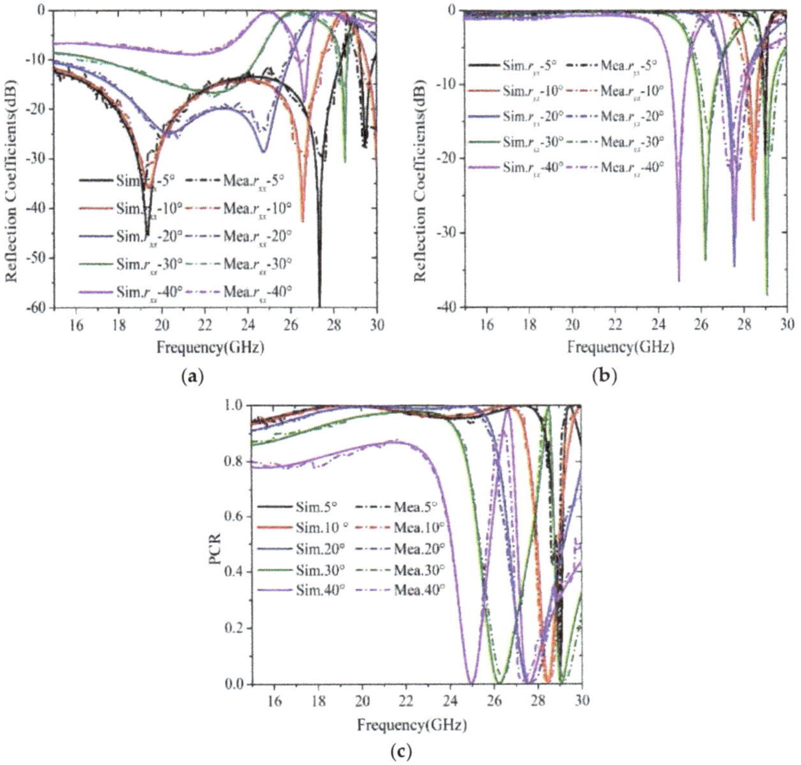

Figure 13. Simulated and measured results for different oblique incident angle. (**a**) Co-polarized reflection coefficient r_{xx}; (**b**) Cross-polarized reflection coefficient r_{yx}; (**c**) PCR.

A comparison between the proposed converter with other polarization converters reported recently is presented in Table 1. As can be seen from Table 1, the proposed polarization converter presents a wider bandwidth compared with those reported in Refs. [25–27] and Refs. [31–36]. Although there are a few designs [28–30] that outperform the bandwidth of the one proposed in this work, these structures are developed using via connection or multi-layer dielectric substrates, and only operates efficiently with normal incidence. The presented structure can provide a good angular stability for the oblique incidence angle in Refs. [32–34], but the bandwidth is not wide enough. Both types of broadband design are obtained in Refs. [35,36], but the angular stability is not improved. From the comparison, it can be concluded that good angular stability with satisfactory bandwidth have been realized using the design proposed in this work.

As a matter of fact, it is always a challenge to create large angle stability with a broad bandwidth. One reason is that broadband polarization conversion requires multi-resonances to ensure enough bandwidth. However, for large angle incidence, higher order resonances cannot be always ensured. Therefore, one way is to minimize the unit cell using a folded patten. However, this will introduce difficulty to the realization of low order resonance. To overcome this difficulty, one may cut slots to prolong the current path so that low order resonance may be obtained. However, if larger angle stability and wider bandwidth are required, these techniques may not be sufficient. Therefore, more techniques may need further investigation.

Table 1. Performance comparison of polarization converters.

Ref.	Frequency (GHz)	Bandwidth	PCR	Thickness	Metallic Layers	Angle Stability
[25]	4.4–5.3, 9.45–13.6	18.56%, 36.01%	90%	$0.044\lambda_{max}$	2	0°
[26]	7.74–14.44	60.41%	90%	$0.077\lambda_{max}$	2	0°
[27]	8.2–23	94.87%	80%	$0.082\lambda_{max}$	2	0°
[28]	6.7–23.4	110.96	90%	$0.073\lambda_{max}$	2	0°
[29]	16.2–57	111.47%	89%	$0.082\lambda_{max}$	2	0°
[30]	7.8–34.7	126.56%	90%	$0.104\lambda_{max}$	3	0°
[31]	10.2–20.5	67%	90%	$0.068\lambda_{max}$	2	30°
[32]	11.5–21.8	61.86%	90%	$0.076\lambda_{max}$	2	40°
[33]	17.97–40.23	76.49%	90%	$0.071\lambda_{max}$	2	40°
[34]	6.36–6.59, 10.54–13.56	3.55%, 25.06%	90%	$0.05\lambda_{max}$	2	45°
[35]	17–42	84.7%	90%	$0.085\lambda_{max}$	2	20°
[36]	4.6–12.27	90.93%	90%	$0.076\lambda_{max}$	2	25°
This work	9.83–29.37	99.69%	90%	$0.082\lambda_{max}$	2	40°

5. Conclusions

An ultra-broadband reflective polarization converter based on anisotropic metasurface is proposed. The structure converts the linearly polarized incident wave into its orthogonal counterpart over a wide frequency range of 9.83–29.37 GHz. It has been verified that the PCR of the proposed converter is above 90% within the operating band and nearly reaches 100% at three resonant frequencies of 10.68, 18.74 and 28.10 GHz. With surface current distribution analysis, the ultra-broadband polarization conversion is created due to the multiple resonances between top and bottom layers. Moreover, for a wide-angle incidence, this efficiency can be maintained higher than 80%, even though the incident angle reaches 40°. A fabricated prototype is measured, and confirms that the experimental results are in good agreement with the simulation. Compared with other designs in the literature, the proposed polarization converter has a good performance of high efficiency, ultra-broadband and angular stability.

Author Contributions: Conceptualization, X.L.; methodology, B.Z.; software, R.Z. and X.Y.; validation, C.Z.; data curation, B.Z.; writing—original draft preparation, B.Z.; writing—review and editing, X.L.; funding acquisition, X.L., Y.W. and X.Y. All authors have read and agreed to the published version of the manuscript.

Funding: This work is funded in part by the National Natural Science Foundation of China (61871003, 62101005), the Natural Science Foundation of Anhui province (No. 2108085QF256), and Natural Science Research Project for Universities in Anhui Province (KJ2021A0101, KJ2021A1203).

Institutional Review Board Statement: Not applicable.

Informed Consent Statement: Not applicable.

Data Availability Statement: Not applicable.

Conflicts of Interest: The authors declare no conflict of interest.

References

1. Xu, J.; Li, R.; Qin, J.; Wang, S.; Han, T. Ultra-Broadband Wide-Angle Linear Polarization Converter Based on H-Shaped Metasurface. *Opt. Express* **2018**, *26*, 20913. [CrossRef] [PubMed]
2. Yang, Z.; Kou, N.; Yu, S.; Long, F.; Yuan, L.; Ding, Z.; Zhang, Z. Reconfigurable Multifunction Polarization Converter Integrated with PIN Diode. *IEEE Microw. Wirel. Compon. Lett.* **2021**, *31*, 557–560. [CrossRef]
3. Pouyanfar, N.; Nourinia, J.; Ghobadi, C. Multiband and Multifunctional Polarization Converter Using an Asymmetric Metasurface. *Sci. Rep.* **2021**, *11*, 9306. [CrossRef] [PubMed]
4. Zeng, Q.; Ren, W.; Zhao, H.; Xue, Z.; Li, W. Dual-Band Transmission-Type Circular Polariser Based on Frequency Selective Surfaces. *IET Microw. Antennas Propag.* **2019**, *13*, 216–222. [CrossRef]
5. Li, S.; Zhang, X. Asymmetric Tri-Band Linear-to-Circular Polarization Converter in Transmission Mode. *Int. J. RF Microw. Comput.-Aided Eng.* **2020**, *30*, e21959. [CrossRef]

6. Liu, K.; Wang, G.; Cai, T.; Li, T. Dual-Band Transmissive Circular Polarization Generator with High Angular Stability. *Opt. Express* **2020**, *28*, 14995. [CrossRef] [PubMed]
7. Liao, K.; Liu, S.; Zheng, X.; Zhang, X.; Shao, X.; Kong, X.; Hao, Z. A Polarization Converter with Single-Band Linear-to-Linear and Dual-Band Linear-to-Circular Based on Single-Layer Reflective Metasurface. *Int. J. RF Microw. Comput.-Aided Eng.* **2022**, *32*, e22955. [CrossRef]
8. Chen, C.-Y.; Hsu, M.-C.; Hu, C.D.; Lin, Y.C. Natural Negative-Refractive-Index Materials. *Phys. Rev. Lett.* **2021**, *127*, 237401. [CrossRef]
9. Madl, T. Patchy Proteins Form a Perfect Lens. *Science* **2017**, *357*, 546–547. [CrossRef] [PubMed]
10. Zhan, J.; Li, K.; Zhou, Y.; Liu, X.; Ma, Y. Ultrathin Conformal Magnetic Invisible Cloak for Irregular Objects. *ACS Appl. Mater. Interfaces* **2021**, *13*, 17104–17109. [CrossRef]
11. Lin, J.-H.; Yen, T.-J.; Huang, T.-Y. Design of Annulus-Based Dielectric Metamaterial Cloak with Properties of Illusion Optics. *J. Opt.* **2020**, *22*, 085101. [CrossRef]
12. Tuan, T.S.; Hoa, N.T.Q. Numerical Study of an Efficient Broadband Metamaterial Absorber in Visible Light Region. *IEEE Photonics J.* **2019**, *11*, 4600810. [CrossRef]
13. Nguyen, T.Q.H.; Nguyen, T.K.T.; Nguyen, T.Q.M.; Cao, T.N.; Phan, H.L.; Luong, N.M.; Le, D.T.; Bui, X.K.; Truong, C.L.; Vu, D.L. Simple Design of a Wideband and Wide-Angle Reflective Linear Polarization Converter Based on Crescent-Shaped Metamaterial for Ku-Band Applications. *Opt. Commun.* **2021**, *486*, 126773. [CrossRef]
14. Nguyen, T.K.T.; Nguyen, T.M.; Nguyen, H.Q.; Cao, T.N.; Le, D.T.; Bui, X.K.; Bui, S.T.; Truong, C.L.; Vu, D.L.; Nguyen, T.Q.H. Simple Design of Efficient Broadband Multifunctional Polarization Converter for X-Band Applications. *Sci. Rep.* **2021**, *11*, 2032. [CrossRef] [PubMed]
15. Wu, X.; Gao, X.; Tang, L. A High-Performance Transmissive Circular Polarization Converter Based on a Modified Fishnet Metasurface. *Appl. Phys. B* **2021**, *127*, 31. [CrossRef]
16. Wang, L.; Lv, F.; Xiao, Z.; Ma, X. Theoretical and Experimental Verification of 90° Polarization Converter Based on Chiral Metamaterials. *Plasmonics* **2021**, *16*, 199–204. [CrossRef]
17. Zhang, W.; Li, J.-Y.; Xie, J. A Broadband Circular Polarizer Based on Cross-Shaped Composite Frequency Selective Surfaces. *IEEE Trans. Antennas Propag.* **2017**, *65*, 5623–5627. [CrossRef]
18. Couto, M.M.; Silva, M.W.B.; Campos, A.L.P.S. A Novel Ultra-Wideband Reflective Cross-Polarization Converter Based on Anisotropic Metasurface. *J. Electromagn. Waves Appl.* **2021**, *35*, 1652–1662. [CrossRef]
19. Xu, Z.; Sheng, H.; Wang, Q.; Zhou, L.; Shen, Y. Terahertz Broadband Polarization Converter Based on the Double-Split Ring Resonator Metasurface. *SN Appl. Sci.* **2021**, *3*, 763. [CrossRef]
20. Cui, Z.; Xiao, Z.; Chen, M.; Lv, F.; Xu, Q. A Transmissive Linear Polarization and Circular Polarization Cross Polarization Converter Based on All-Dielectric Metasurface. *J. Electron. Mater.* **2021**, *50*, 4207–4214. [CrossRef]
21. Liu, D.; Xiao, Z.; Ma, X.; Wang, Z. Broadband Asymmetric Transmission and Multi-Band 90° Polarization Rotator of Linearly Polarized Wave Based on Multi-Layered Metamaterial. *Opt. Commun.* **2015**, *354*, 272–276. [CrossRef]
22. Naseri, P.; Matos, S.A.; Costa, J.R.; Fernandes, C.A.; Fonseca, N.J.G. Dual-Band Dual Linear to Circular Polarization Converter in Transmission Mode Application to K/Ka-Band Satellite Communications. *IEEE Trans. Antennas Propag.* **2018**, *66*, 7128–7137. [CrossRef]
23. Zhao, J.; Cheng, Y.; Huang, T.; Liu, P. A Dual-band Linear-to-circular Polarization Converter with Robustness under Oblique Incidences. *Microw. Opt. Technol. Lett.* **2021**, *63*, 361–366. [CrossRef]
24. Lin, B.; Lv, L.; Guo, J.; Liu, Z.; Ji, X.; Wu, J. An Ultra-Wideband Reflective Linear-to-Circular Polarization Converter Based on Anisotropic Metasurface. *IEEE Access* **2020**, *8*, 82732–82740. [CrossRef]
25. Huang, X.; Yang, H.; Zhang, D.; Luo, Y. Ultrathin Dual-Band Metasurface Polarization Converter. *IEEE Trans. Antennas Propag.* **2019**, *67*, 4636–4641. [CrossRef]
26. Zheng, Q.; Guo, C.; Vandenbosch, G.A.E.; Yuan, P.; Ding, J. Dual-broadband Highly Efficient Reflective Multi-polarisation Converter Based on Multi-order Plasmon Resonant Metasurface. *IET Microw. Antennas Propag.* **2020**, *14*, 967–972. [CrossRef]
27. Li, F.; Chen, H.; Zhang, L.; Zhou, Y.; Xie, J.; Deng, L.; Harris, V.G. Compact High-Efficiency Broadband Metamaterial Polarizing Reflector at Microwave Frequencies. *IEEE Trans. Microw. Theory Tech.* **2019**, *67*, 606–614. [CrossRef]
28. Zaker, R.; Sadeghzadeh, A. A Low-Profile Design of Polarization Rotation Reflective Surface for Wideband RCS Reduction. *IEEE Antennas Wirel. Propag. Lett.* **2019**, *18*, 1794–1798. [CrossRef]
29. Omar, A.A.; Hong, W.; Al-Awamry, A.; Mahmoud, A.-E. A Single-Layer Vialess Wideband Reflective Polarization Rotator Utilizing Perforated Holes. *IEEE Antennas Wirel. Propag. Lett.* **2020**, *19*, 2053–2056. [CrossRef]
30. Jia, Y.; Liu, Y.; Zhang, W.; Gong, S. Ultra-Wideband and High-Efficiency Polarization Rotator Based on Metasurface. *Appl. Phys. Lett.* **2016**, *109*, 051901. [CrossRef]
31. Karamirad, M.; Ghobadi, C.; Nourinia, J. Metasurfaces for Wideband and Efficient Polarization Rotation. *IEEE Trans. Antennas Propag.* **2021**, *69*, 1799–1804. [CrossRef]
32. Liu, Z.; Zhao, B.; Jiao, C.; Zhao, L.; Han, X. Broadband Cross-Polarization Conversion Metasurface Based on Cross-Shaped Resonators. *Appl. Phys. A* **2021**, *127*, 825. [CrossRef]

33. Shukoor, M.A.; Dey, S.; Koul, S.K.; Poddar, A.K.; Rohde, U.L. Broadband linear-cross and circular-circular Polarizers with Minimal Bandwidth Reduction at Higher Oblique Angles for RCS Applications. *Int. J. RF Microw. Comput.-Aided Eng.* **2021**, *31*, e22693. [CrossRef]
34. Salman, M.S.; Khan, M.I.; Tahir, F.A.; Rmili, H. Multifunctional Single Layer Metasurface Based on Hexagonal Split Ring Resonator. *IEEE Access* **2020**, *8*, 28054–28063. [CrossRef]
35. Yang, X.; Qi, T.; Zeng, Y.; Liu, X.; Lu, G.; Cai, Q. Broadband Reflective Polarization Rotator Built on Single Substrate. *Electronics* **2021**, *10*, 916. [CrossRef]
36. Wu, L.-P.; Fang, Q.-H.; Pan, W.-K.; Li, M.-H.; Dong, J.-F. Realizing of Cross-Polarization Conversion for Linearly Polarized Waves in Multiband by Orthotropic Metasurface. *J. Electromagn. Waves Appl.* **2022**, *36*, 101–114. [CrossRef]
37. Liu, X.; Wang, H.; Yang, X.; Wang, J. Quad-Band Circular Polarized Antenna for GNSS, 5G and WIFI-6E Applications. *Electronics* **2022**, *11*, 1133. [CrossRef]
38. Gao, X.; Han, X.; Cao, W.-P.; Li, H.O.; Ma, H.F.; Cui, T.J. Ultrawideband and High-Efficiency Linear Polarization Converter Based on Double V-Shaped Metasurface. *IEEE Trans. Antennas Propagat.* **2015**, *63*, 3522–3530. [CrossRef]

electronics

MDPI

Article

Design and Analysis of Polarization-Independent, Wide-Angle, Broadband Metasurface Absorber Using Resistor-Loaded Split-Ring Resonators

Abdulrahman Ahmed Ghaleb Amer [1], Syarfa Zahirah Sapuan [1,*], Abdullah Alzahrani [2], Nasimuddin Nasimuddin [3], Ali Ahmed Salem [4,*] and Sherif S. M. Ghoneim [2]

1 Faculty of Electrical and Electronic Engineering, Universiti Tun Hussein Onn Malaysia (UTHM), Parit Raja 86400, Johor, Malaysia; aag2014ye@gmail.com
2 Department of Electrical Engineering, College of Engineering, Taif University, P.O. Box 11099, Taif 21944, Saudi Arabia; aatyah@tu.edu.sa (A.A.); s.ghoneim@tu.edu.sa (S.S.M.G.)
3 Institute for Infocomm Research, A-STAR, Singapore 138632, Singapore; nasimuddin@i2r.a-star.edu.sg
4 Institute of High Voltage and High Current, School of Electrical Engineering, Universiti Teknologi Malaysia, Johor Bahru 81310, Johor, Malaysia
* Correspondence: syarfa@uthm.edu.my (S.Z.S.); ahmedali.a@utm.my (A.A.S.)

Abstract: Metasurface (MS) absorbers with polarization-insensitivity and wide-angle reception features have attracted much attention due to their unique absorption property. A polarization-insensitive broadband MS absorber structure, having wide-angle reception based on square split-ring resonators (SSRRs) and loaded with lumped resistors, is proposed in this paper. The proposed MS unit cell consists of a fixed-thickness FR4 dielectric substrate and a variable air-thickness substrate. The simulation results show that the proposed MS absorber is stable across a wide angular range for both normal and oblique incidences. Furthermore, the simulated results show that some parameters, such as unit-cell geometry and lumped resistors, can be varied to improve the performance of the MS absorber. The experimental results indicate that the proposed MS absorber can be achieved an absorption higher than 90% across the frequency range from 1.89 GHz to 6.85 GHz with a relative bandwidth of 113%, which is in agreement with simulation results. Thus, the proposed MS absorber can be more suitable in RF energy harvesting or wireless power transfer applications.

Keywords: metasurface absorber; broadband; split-ring resonator; polarization-independent; wide angle

Citation: Amer, A.A.G.; Sapuan, S.Z.; Alzahrani, A.; Nasimuddin, N.; Salem, A.A.; Ghoneim, S.S.M. Design and Analysis of Polarization-Independent, Wide-Angle, Broadband Metasurface Absorber Using Resistor-Loaded Split-Ring Resonators. *Electronics* **2022**, *11*, 1986. https://doi.org/10.3390/electronics11131986

Academic Editors: Naser Ojaroudi Parchin, Mohammad Ojaroudi and Raed A. Abd-Alhameed

Received: 2 June 2022
Accepted: 19 June 2022
Published: 24 June 2022

1. Introduction

Electromagnetic (EM) absorbers are used to convert energy in EM waves into other forms of energy and in a variety of applications, such as EM interfaces, EM compatibility [1,2], and anechoic chambers [3]. For EM absorber applications, one can use absorbing structures such as a Jaumann absorber [4], a wedge-tapered absorber [5], or plasma [6]. However, their use is difficult and costly. Metamaterial absorbers (MMA) are proposed to overcome these challenges. Metamaterials can be defined as an artificial structure that offers startling features such as negative permeability and/or permittivity, which results in a negative or positive refractive index and backward propagation [7]. The metamaterial perfect absorber (MPA) was first demonstrated by Landy, et al. in 2008 [8], which has received more attention because of its ability to absorb EM waves with a near-unity absorption efficiency. The MMA can be formed by an ensemble of periodic subwavelength resonating unit cells with impedance matching that of the free space. Such properties of MPA prompted researchers to use it in a variety of applications, including undesired frequency absorption [9], energy harvesting [10–12], sensing, and optics [13,14]. The MPA prefers to be polarization-independent, broadband, and has wide-angle reception in addition to achieving a near-unity absorption.

However, MMA has a limited bandwidth due to its resonating characteristic. The MMA can be designed through a variety of approaches to enhance its bandwidth using multilayer structure [15,16] and resistor film [17,18]. However, because of the difficulty of impedance matching to free space for each resonance frequency, the MMAs based on multilayer structures would suffer from a lack of absorption level as bandwidth increased. In addition, the MMAs with lumped elements loaded on the top metallic resonator with the ground plane have been demonstrated to achieve broadband absorption [19–24]. For example, a broadband MMA with a resistor element was designed in [19], where a wider bandwidth of about 1.5 GHz with 90% absorption was obtained within a frequency range from 3.8 GHz to 5.3 GHz. In addition, the MMA with resistor elements was presented to demonstrate a wider bandwidth within a frequency range from 2.85 GHz up to 5.31 GHz [20]. In [21], a broadband absorber was designed using a dual-layer MS and welded with the resistor elements to show a wider bandwidth from 3.8 GHz to 14.8 GHz. A broadband MMA comprised of multilayer resonators separated by two dielectric layers is designed in [22]. The resistive elements are welded on the top metallic resonator to achieve a wider bandwidth between 4 GHz and 16 GHz. This method requires more resistor loads and is complex for a large fabrication array. In [23], a broadband MMA with four resistor elements was designed to have a wider bandwidth across frequency band from 8 GHz up to 18 GHz. The four metallic holes were used to improve the absorption performance, making fabrication more complicated. In order to efficiently absorb EM power at normal incidence, a broadband MMA was designed to show a wider bandwidth across a frequency range from 4.4 GHz to 18 GHz [24]. The proposed MMA comprises multi resonator slabs loaded with lumped resistors, making fabrication more complex and expensive. In another approach, to produce broadband with a higher absorption level, a MMA with an air layer sandwiched between the dielectric substrate and a metal ground was used [24,25]. In [25], an MMA consisting of a 16-sided equilateral shape hosted on the FR4 substrate was proposed. A wider bandwidth of more than 2 GHz was observed within a frequency range from 1.35 GHz to 3.5 GHz. In addition, Q. Wang, et al., have realized a broadband MMA using a meander wire structure that operates within 1.91 GHz to 4.24 GHz [26]. Therefore, researchers have been developed single, multiband, broadband, and flexible metamaterial absorbers [26–32]. In [30], an absorber with fractional bandwidth of about 91.67% was designed. Similarly, the absorbers with expensive flexible substrate materials such as polyethylene terephthalate (PET), polydimethylsiloxane (PDMS), and polyimide (PI) were proposed to increase the fractional bandwidth [31–33]. However, the design of a flexible absorber requires additional fabrication processes and costs. Furthermore, broadband remains highly desirable for practical applications, especially at the lower frequency range.

This work presents a broadband MS absorber based on the square split-ring resonator (SSRR) loaded with lumped resistors. The proposed MS resonator is hosted on an FR4 dielectric substrate and is followed by a full-copper ground plane. An air layer with variable thickness is sandwiched between the FR4 substrate and the ground plane. The proposed structure-based approach is low cost with essay fabrication. Simulation results show that the MS absorber has more than 90% absorption efficiency across the frequency range from 1.88 GHz to 6.4 GHz. Furthermore, higher absorption can be achieved at a normal and oblique incidence up to 60° for both TE and TM polarizations. The electric and surface current distributions are investigated to explain the absorption mechanism. The polarization-insensitivity and wide-incident absorption properties of TE and TM polarization waves are numerically validated. Furthermore, the simulation results show that the critical parameters influence absorption peaks, strength, and bandwidth. Finally, the proposed MS absorber's performance was then verified experimentally.

2. Metasurface Absorber Design

Figure 1 depicts the proposed MS unit-cell absorber. As shown in Figure 1b, the proposed unit cell comprises three dielectric layers: top (FR4), middle (air), and bottom (FR4). Figure 1a depicts the SSRR hosted on the top FR-4 substrate ($\varepsilon_r = 4.3$ and $\delta = 0.025$).

A full copper plate (thickness = 35 μm and conductivity = 5.8 × 10^{-7} s/m) covers the top side of the bottom FR4 substrate to block the transient. The top and bottom FR4 substrates have a uniform thickness (1.6 mm), but the middle air layer thickness (t) is variable. Four chip resistors are welded on the splits of the resonator to achieve a broadband MS absorber. The geometrical parameters shown in Figure 1c, determined for the simulation, are: P = 40 mm, L1 = 25.8 mm, L2 = 31.53 mm, W1 = 1.4 mm, W2 = 1 mm, S = 7.1 mm, g = 8 mm, and t = 14 mm. The four lumped resistors have the same resistance value of R = 560 Ω. Numerical simulations were performed in the CST Microwave Studio using the frequency domain solver to demonstrate the performance of the proposed MS absorber and gain insight into its optimized microwave absorption mechanism. The proposed MS absorber comprises a periodically arranged unit cell so that the performance of the central unit cell is investigated instead of calculating the overall performance of the whole MS absorber footprint. For the simulation, the periodic boundaries are applied with an electric field (E-field) along the *x*-axis and a magnetic field (H-field) along the *y*-axis to model an infinite array, as seen in Figure 2. Floquet ports are applied along the *z*-axis to model a linearly polarized incident wave.

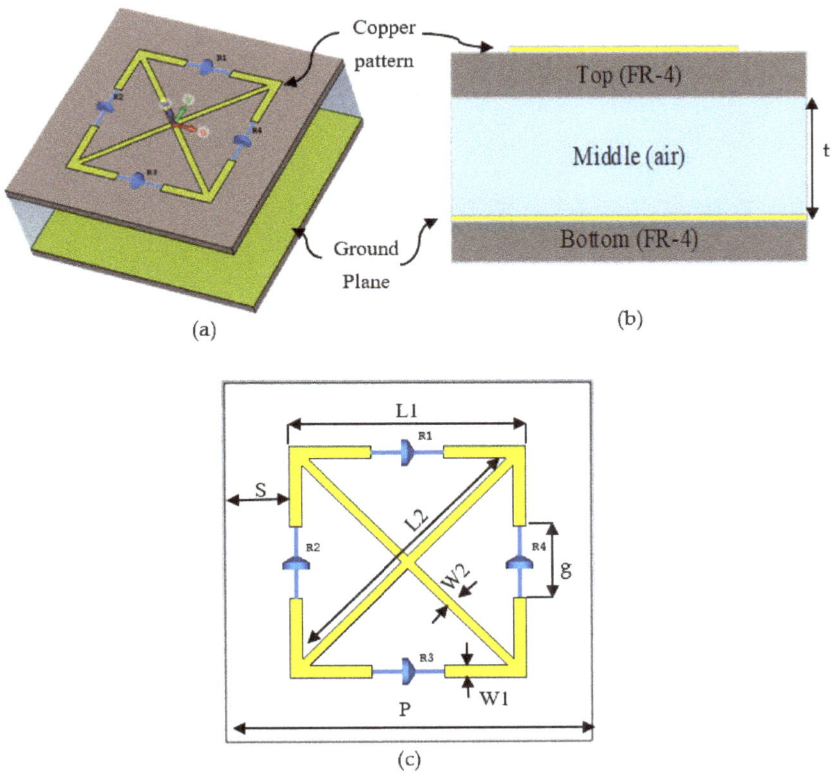

Figure 1. Geometry of proposed MS absorber: (**a**) perspective view, (**b**) side view, and (**c**) top view.

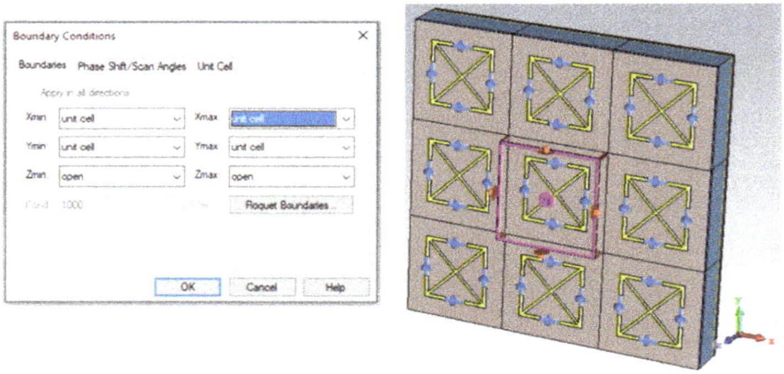

Figure 2. Boundary setup in CST.

The effective input impedance of the MS absorber in terms of frequency can be calculated as in Equation (1) [34]

$$Z_{in}(\omega) = \sqrt{\frac{(1 + S_{11}(\omega))^2 - S_{21}^2(\omega)}{(1 - S_{11}(\omega))^2 - S_{21}^2(\omega)}}, \text{ where } \omega = 2\pi f \tag{1}$$

Hence, to achieve a perfect absorption, the effective input impedance should match with the free space $Z_{in}(\omega) = Z_o$, which can be achieved by adjusting the MS structure's permittivity and permeability. Thus, the absorption response $A(\omega)$ can be calculated using Equation (2), as given below [8]

$$A(\omega) = 1 - R(\omega) - T(\omega) \tag{2}$$

where, $R(\omega) = |S_{11}|^2$ and $T(\omega) = |S_{21}|^2$ represent reflection and transmission coefficients, respectively. The perfect absorption requires minimal reflection and transmission. Because the bottom layer of the MS structure is made of copper, the transmitted wave is negligibly small.

3. Results and Discussion

Figure 3 shows the simulated absorption spectra for the proposed MS absorber under various conditions. Figure 3a shows the simulated absorption and reflection coefficients of the proposed MS absorber under normal incidence, where EM waves fall onto the structure with a parallel E-field and a perpendicular H-field. As shown in Figure 3a, the proposed MS absorber offers absorption of around 90% across the frequency range from 1.88 GHz to 6.4 GHz, and lower reflection is obtained as well. Therefore, the absorption magnitude is calculated from the reflection coefficient only since the transmission coefficient is almost zero. Furthermore, the absorption for two different substrate conditions is shown in Figure 3b. The results show that the absorption spectra of the dielectric substrate (FR-4) under loss and loss-free conditions are nearly identical. In addition, the simulated absorption for the MS structure with and without loading resistors is shown in Figure 3c. When the MS absorber is not loaded with resistors, the absorptions are around 21.0%, 81.7%, and 96.0% at 1.8 GHz, 6.63 GHz, and 7.0 GHz, respectively. Results show that lumped resistors and the MS structure's resonance response are the primary factors of the MS absorber's broadband absorption.

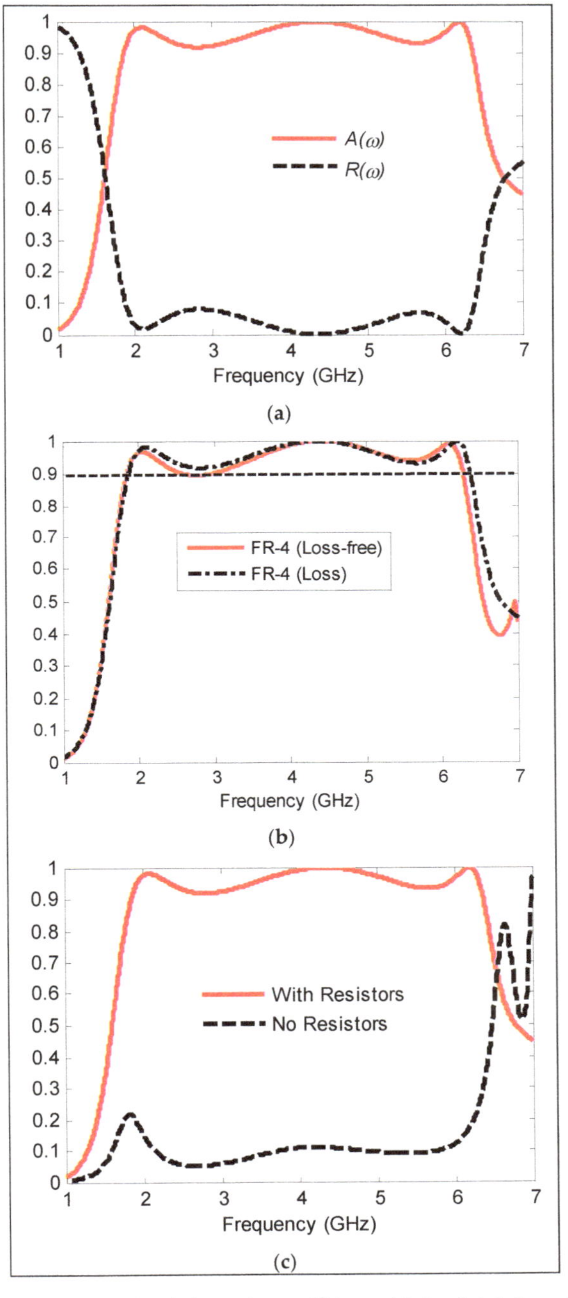

Figure 3. Simulated absorption coefficients: (**a**) simulated absorption and reflection under normal incidence, (**b**) absorption under two different loss conditions, and (**c**) the absorption of the MS absorber with and without resistor loads.

The MS absorber's performance is studied using E-field and surface current distributions at 2.09 GHz, 4.34 GHz, and 6.2 GHz to reveal the physical mechanism of the MS structure. These frequencies are chosen because the highest absorption levels of about 98.2%, 99.9%, and 99.8% are achieved at these frequencies of 2.09 GHz, 4.34 GHz, and 6.2 GHz, respectively. Figure 4 depicts the proposed MS absorber's E-field distribution. At 2.09 GHz, the E-field is distributed throughout the resonator's upper and lower sides and is more concentrated along the splits, as shown in Figure 4a. Furthermore, the power density at the frequency of 2.09 GHz is dissipated at the top and bottom resistor elements. At 4.34 GHz, the E-field distribution on the resonator's right and left sides are observed; these concentrate more on the split's left and right edges, as shown in Figure 4b. The resonator's induced power is wasted on the left and right resistors. At 6.2 GHz, the E-field is distributed along the external and internal resonators and is more concentrated at the external resonator's edges, as shown in Figure 4c. In addition, the maximum induced power density on the surface is dissipated on four resistor loads at 6.2 GHz, resulting in the highest absorption level.

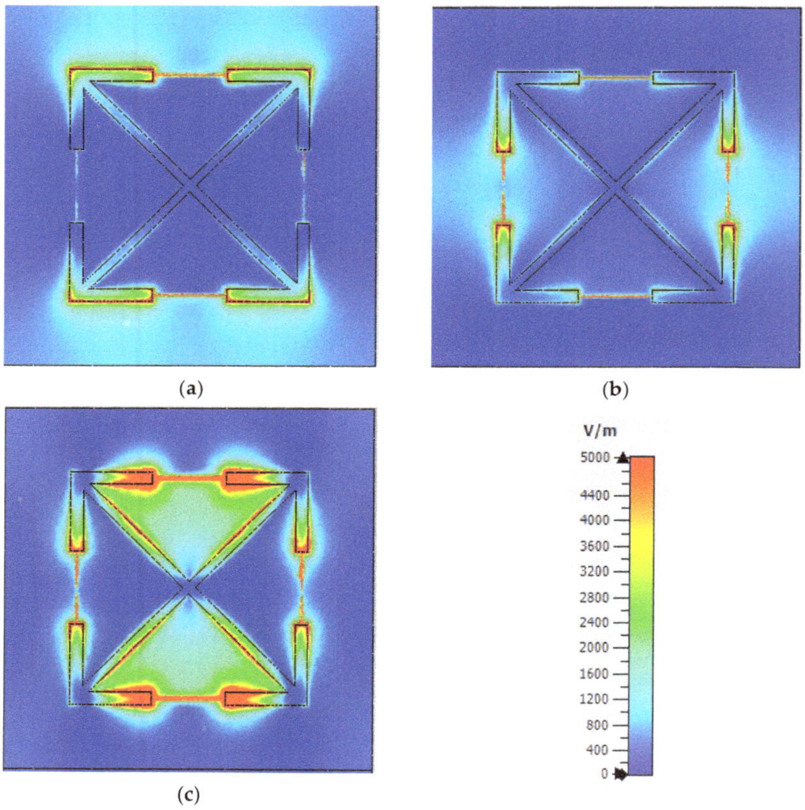

Figure 4. E-field distributions: (**a**) 2.09 GHz, (**b**) 4.34 GHz, and (**c**) 6.2 GHz.

The proposed MS absorber's surface current distribution was anti-parallel, resulting in a significant magnetic resonance and a higher absorption, as shown in Figure 5. Figure 5a shows the dominant surface current distribution at the inner traces at 2.09 GHz. Figure 5b shows a highly distributed surface current at 4.34 GHz with parallel circulation currents at the outer trace. In addition, the surface current distribution along the proposed structure is responsible for the 6.2 GHz resonance, as shown in Figure 5c.

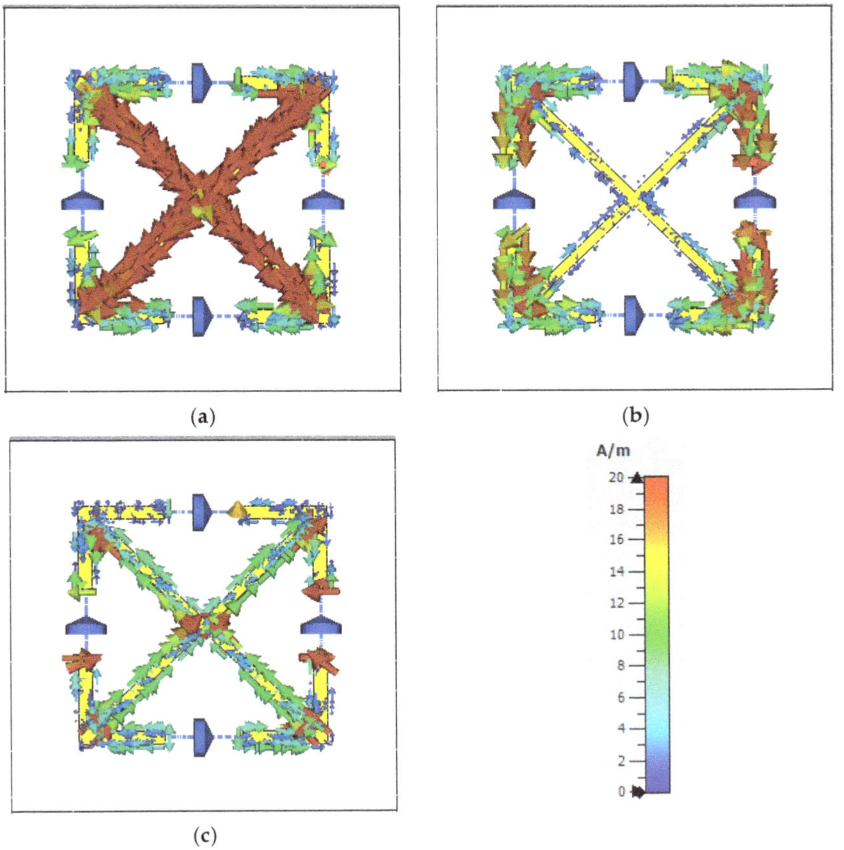

Figure 5. Surface current distributions: (**a**) 2.09 GHz, (**b**) 4.34 GHz, and (**c**) 6.2 GHz.

3.1. Absorption Mechanisms and Different Design Parameters

To investigate the relationship between design parameters and absorption responses, five major geometric parameters are numerically computed and analyzed: air layer thickness (t), external resonator's width (W1), inner resonator's width (W2), split's size (g), and lumped resistance (R). Figure 6a–e depicts the absorption spectra for five different parameters (t, W1, W2, g, and R), where only one parameter is changed simultaneously, keeping the others constant.

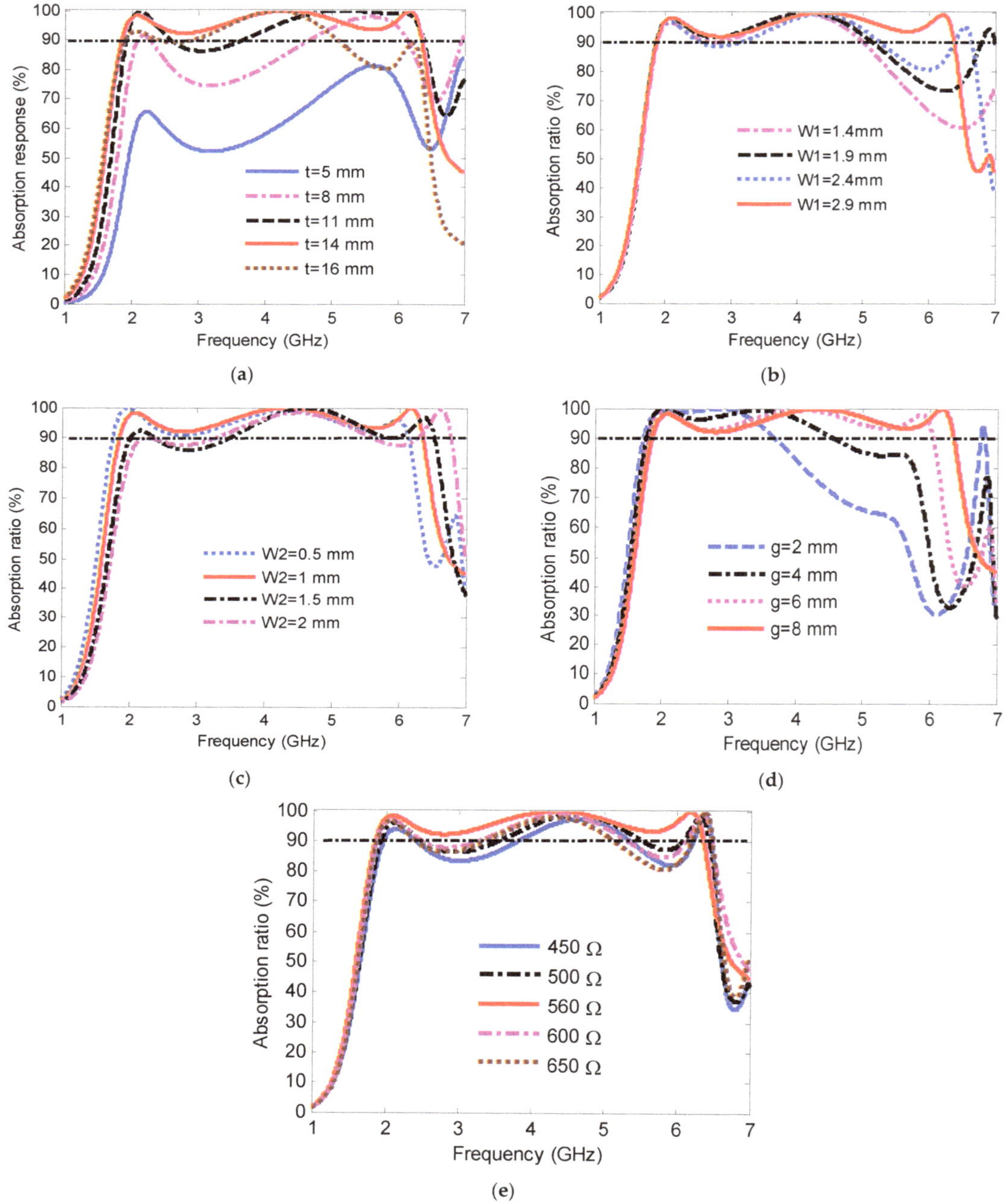

Figure 6. Simulated absorption spectra of the MS absorber: (**a**) air layer thickness, (**b**) outer resonator's width, (**c**) inner resonator's width, (**d**) split's size, and (**e**) load resistor.

Figure 6a presents the absorption spectra with variation in the air-layer thickness (t) from 5 mm to 16 mm. As shown in Figure 6a, when t changes from 5 mm to 16 mm, the absorption peaks shift to lower frequency, and the absorption coefficient increases to more than 90%. When $t = 14$ mm, the absorbance reaches around 90% across the frequency range from 1.88 to 6.4 GHz. By adding the air layer, the total thickness increases, the dielectric's effective permittivity decreases, and the quality factor Q decreases. Thus, the bandwidth increases with a decrease in the quality factor, as described in Equation (3) [31]

$$BW = \frac{f_o}{Q} = \frac{R}{2\pi L} \tag{3}$$

where f_o = frequency, Q = the quality factor, R = resistance and L = inductance. As shown in Figure 6b, the simulated absorption is presented when the external resonator's width (W1) varies from 1.4 mm to 2.9 mm with steps of 0.5 mm. The absorbance drops to less than 90% around 5.1 GHz when W1 changes from 1.4 mm to 2.4 mm. When W1 = 2.9 mm, a wider bandwidth of about 4.5 GHz is achieved across a frequency range from 1.88 to 6.4 GHz, as shown in Figure 6b. Figure 6c shows the simulated absorption curves when the inner cross-resonator's width (W2) is varied from 0.5 mm to 2 mm in steps of 0.5 mm. As W2 increases, the resonance peaks shift to the right, and the absorption falls below 90%. When the W2 = 1 mm, the maximum absorption value is achieved. With the change of split's size (g) from 2 mm to 8 mm in steps of 2 mm, the absorption curves are shown in Figure 6d. Obviously, with an absorption level higher than 90%, the wider bandwidth of the absorption is achieved as g increases. The maximum absorption value is observed when g = 8 mm. Finally, Figure 6e depicts the absorption level as the resistance is varied from 450 Ω to 650 Ω. There are additional ohmic losses when resistor loads are added to a proposed MS structure. The bandwidth is proportional to the resistor load as described in Equation (3). It is clear that the widest continuous bandwidth with more than 90% absorption is observed when R = 560 Ω. When other resistances are selected, the absorption peaks drop below 90% for frequencies ranging from 2.47 GHz to 3.6 GHz and 5.1 GHz to 6.2 GHz. Thus, a good match between the impedance of the MS structure and free space (377 Ω) can be achieved when the resistor is selected to be R = 560 Ω; this results in a perfect broadband absorption. The above results show that the unit-geometric cell's parameters significantly impact the absorbing capacity and that the lumped resistor plays an important role in improving the performance of the MS.

3.2. Absorption Mechanism at Different Polarization and Incident Angles

In practice, the direction of the EM wave is unknown. So that the design of the MS absorber capable of collecting ambient EM energy with characteristics of polarization insensitivity and a wide incident angle from 0° up to 60° is desirable. Numerical simulation was conducted to investigate the MS absorber's polarization and oblique incident angle properties. Figure 7a,b show the simulated absorption of the MS absorber when the polarization angle (ϕ) changes from 0° up to 180° in the step of 30° for TE- and TM-polarization. The MS structure's symmetrical design results in similar absorption responses for the TE- and TM-polarizations. As shown in Figure 7, the proposed MS absorber provides stable performance, confirming that it could deal with the vertical, horizontal, and circular polarization signals.

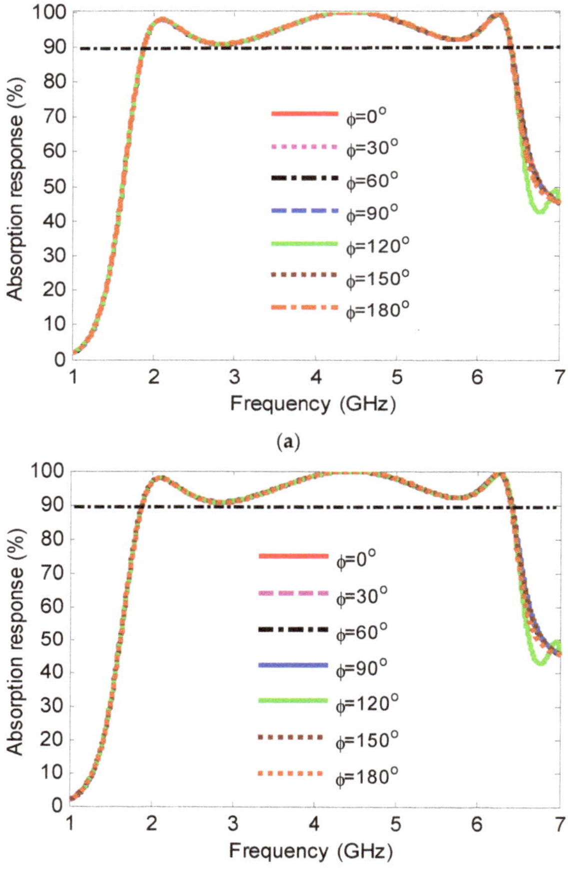

Figure 7. Simulated absorption ratio for various polarization angles: (**a**) TE-polarized and (**b**) TM-polarized.

Figure 8a,b shows the simulated absorption responses for TE-polarized and TM-polarized incident angles, respectively. For TE-polarized incident angles, an absorption ratio of about 80% is achieved with various incident angles up to 45° at a wider frequency range, as shown in Figure 8a. At angle of 60°, the absorption ratio drops below 80% for the frequency band from 2.19 GHz to 3.9 GHz. For TM-polarized incident angles, the absorption ratio remains steady up to 30°, as shown in Figure 8b. When the incident angle increases above of 30°, the absorption ratio of about 80% is achieved. At the frequency band from 4.43 GHz up to 4.7 GHz, the absorption ratio drops to 65%. Based on the above explanation, the absorption responses decreased with increasing incidence angles and decreasing incident magnetic flux between the top and bottom layers.

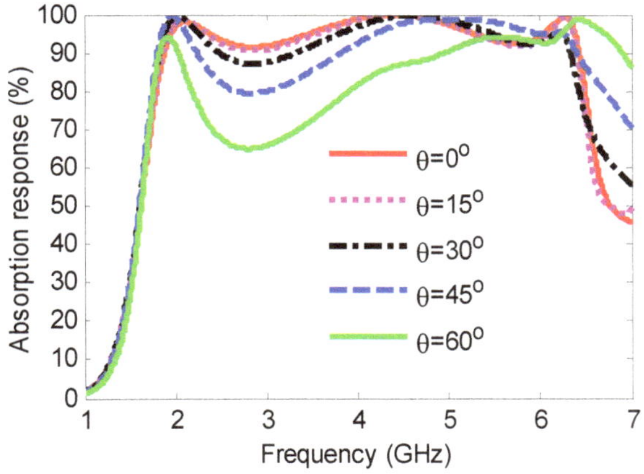

(a)

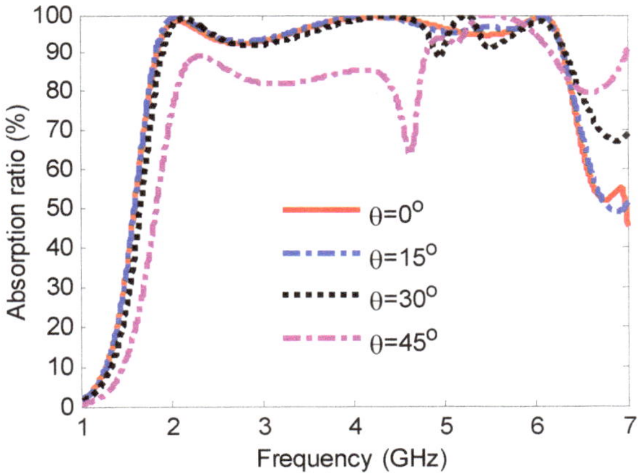

(b)

Figure 8. Simulated absorption ratio for various incident angles: (**a**) TE-polarized and (**b**) TM-polarized.

4. Measurement Verification

Measurement verification of the MS absorber was achieved by prototyping a structure of 7 × 7 unit cells with an overall dimension of 280 mm × 280 mm. Figure 9 shows the MS prototyped structure using PCB technology and measurement setup. The MS resonator is printed on the top of the FR4 layer, where each resonator is connected with four resistor loads to achieve broadband, as shown in Figure 9a. The full copper plate is covered the top view of the bottom FR4 to serve as a ground plane. The top and bottom FR4 substrates are physically connected and supported using four dielectric plastic screws. Full-wave simulation using CST MWS demonstrates that the plastic screws do not affect on the reflection coefficient. The measurement was conducted using two horn antennas (type HF906) as transmitter and receiver, which are then connected to the E5071C vector network analyzer (VNA) ports, as shown in Figure 9b. Tapered wedge absorbers surround the test MS absorber sample to remove the unwanted reflection from the surrounding area/environment. The horn antenna was placed 1 m away from the MS absorber to achieve the maximum radiation power [29]. The measured absorptivity of the proposed absorber was calculated from the reflection coefficient. To begin, reference measurements are taken on a metal surface of the size of the proposed absorber to normalize the measurement results. Then, the fabricated sample is positioned, and the reflection coefficient is measured. The difference between reflected powers from the metal and the fabricated sample is the actual reflection coefficient of the proposed MS absorber. The simulated and measured absorptivity of the fabricated MS absorber sample is shown in Figure 10. It is clear that the measured absorptivity agrees reasonably well with the simulation result except for the small discrepancy caused by the tolerance of the fabrication, especially the height of the air layer and measurement setup. The measured absorptivity above 90% is achieved at the frequency range from 1.89 GHz to 6.85 GHz, and the relative bandwidth is about 113.5%.

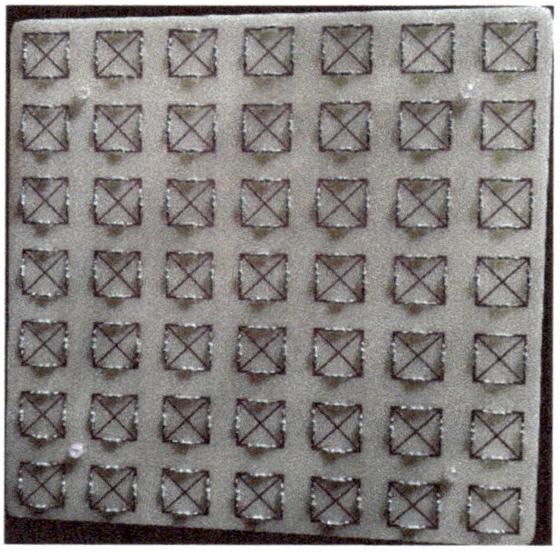

(**a**)

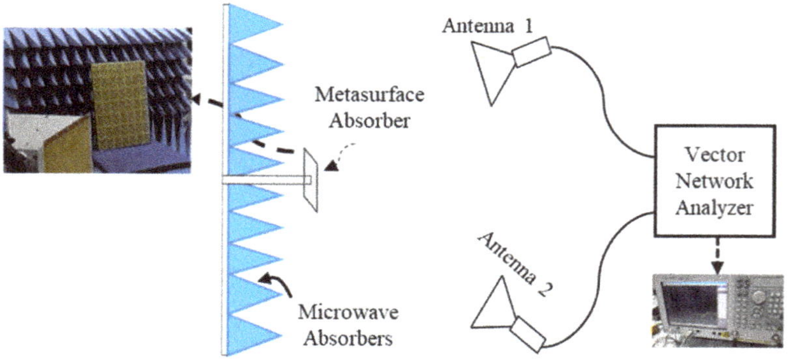

(**b**)

Figure 9. (**a**) Fabricated prototyped of MS absorber and (**b**) measurement setup.

The proposed MS absorber is compared with previously published MMAs in terms of the total thickness, fractional bandwidth, maximum polarization, incident angles, and resistor loads. The proposed MS absorber, as shown in Table 1, has a fractional bandwidth of 113% and provides a higher absorbance for various polarization and incident angles. Furthermore, the proposed MS absorber has fewer resistor loads and a simple structure with easy fabrication.

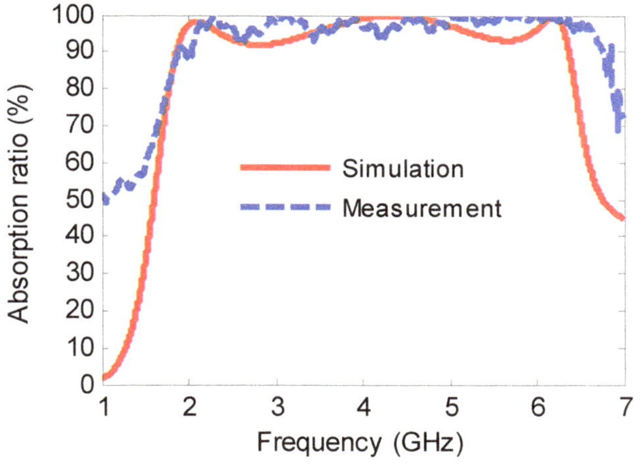

Figure 10. Measured and simulated absorption coefficient.

Table 1. Comparison with other wideband MS absorbers having air layer and loaded with lumped resistors.

Refs.	F_L (GHz)	Total Thickness	Fractional BW (%)	Max Polarized Angle	Max Incident Angle	Resistor No
[25]	1.35	(0.097 λ_L)	88	45	60	8
[26]	1.84	(0.071 λ_L)	105.6	75	50	4
[30]	3.9	(0.098 λ_L)	91.6	90	60	4
[35]	0.86	(0.062 λ_L)	16.4	30	30	4
[36]	0.8	(0.071 λ_L)	108.5	30	30	8
This work	1.88	(0.097 λ_L)	113	180	60	4

5. Conclusions

A broadband MS absorber in the lower frequency range has been demonstrated experimentally for RF energy harvesting and wireless power transfer applications. It was designed based on the square resonator with four splits loaded with lumped resistors to absorb a wide frequency range efficiently. The proposed MS absorber is polarization-insensitivity and shows a broadband absorption for an oblique incidence angle up to 60° for both TE and TM polarizations. Further simulations reveal that the MS structure and lumped resistors have the optimized geometric parameters for achieving the highest absorption coefficient and the widest bandwidth. To better understand the absorption mechanism, E-field and surface current distributions have been presented. Finally, the proposed MS absorber was fabricated and tested to verify the simulated results. Experimental results show that the proposed MS absorber achieves an absorption spectrum of more than 90% in the frequency band from 1.89 GHz up to 6.85 GHz, with the widest fractional bandwidth of up to 113%. Such a broadband MS absorber may be used in different applications such as 5G new radio (5G NR)/EM harvesting, stealth technology, etc.

Author Contributions: Conceptualization, A.A.G.A. and S.Z.S.; methodology, A.A.G.A., S.Z.S., N.N., A.A., A.A.S. and S.S.M.G.; software, A.A.G.A. and S.Z.S.; validation, A.A.G.A., S.Z.S., N.N. and A.A.; formal analysis, A.A.G.A., S.Z.S. and S.S.M.G.; investigation, A.A.G.A., S.Z.S. and N.N.; resources, A.A.G.A. and S.S.M.G.; data curation, A.A.G.A., S.Z.S. and N.N.; writing—original draft preparation, A.A.G.A.; writing—review and editing, A.A.G.A., S.Z.S., A.A., N.N. and S.S.M.G.;

visualization, A.A.G.A., S.Z.S. and N.N.; supervision, S.Z.S.; project administration, S.Z.S., N.N. and A.A.S.; funding acquisition, A.A.G.A. and S.Z.S. All authors have read and agreed to the published version of the manuscript.

Funding: This work is supported by Taif university Researchers Supporting Project TURSP-2020/34, Taif University, Taif, Saudi Arabia and Universiti Tun Hussein Onn Malaysia Publisher's Office via Publication Fund E15216.

Acknowledgments: The authors thank Taif University Researchers Supporting Project TURSP-2020/34, Taif University, Taif, Saudi Arabia for supporting this work. Furthermore, Communication of this research is made possible through monetary assistance by Universiti Tun Hussein Onn Malaysia and the UTHM Publisher's Office via Publication Fund E15216 and Universiti Teknologi Malaysia (UTM) for the Post-Doctoral Fellowship Scheme under the Professional Development Research University Grant (05E68).

Conflicts of Interest: The authors declare no conflict of interest.

References

1. Shukoor, M.A.; Dey, S.; Koul, S.K. A Simple Polarization-Insensitive and Wide Angular Stable Circular Ring Based Undeca-Band Absorber for EMI/EMC Applications. *IEEE Trans. Electromagn. Compat.* **2021**, *63*, 1025–1034. [CrossRef]
2. Munaga, P.; Ghosh, S.; Bhattacharyya, S.; Chaurasiya, D.; Srivastava, K.V. An Ultra-thin Dual-Band Polarization-Independent Metamaterial Absorber for EMI/EMC Applications. In Proceedings of the 2015 9th European Conference on Antennas and Propagation (EuCAP), Lisbon, Portugal, 13–17 April 2015; Volume 2, pp. 11–14.
3. Chung, B.K.; Chuah, H.T. Modeling of RF absorber for application in the design of anechoic chamber. *Prog. Electromagn. Res.* **2003**, *43*, 273–285. [CrossRef]
4. Kazemzadeh, A.; Karlsson, A. Multilayered wideband absorbers for oblique angle of incidence. *IEEE Trans. Antennas Propag.* **2010**, *58*, 3637–3646. [CrossRef]
5. Bucci, O.M.; Franceschetti, G. Scattering from Wedge-Tapered Absorbers. *IEEE Trans. Antennas Propag.* **1971**, *19*, 96–104. [CrossRef]
6. Ha, J.; Shin, W.; Lee, J.H.; Kim, Y.; Kim, D.; Lee, Y.; Yook, J.G. Effect of plasma area on frequency of monostatic radar cross section reduction. *J. Electromagn. Eng. Sci.* **2017**, *17*, 153–158. [CrossRef]
7. Min Zhong, M.Z. Influence of dielectric layer on negative refractive index and transmission of metal-dielectric-metal sandwiched metamaterials. *Chin. Opt. Lett.* **2014**, *12*, 041601–041603. [CrossRef]
8. Landy, N.I.; Sajuyigbe, S.; Mock, J.J.; Smith, D.R.; Padilla, W.J. Perfect metamaterial absorber. *Phys. Rev. Lett.* **2008**, *100*, 207402. [CrossRef]
9. Faruque, M.R.I.; Rahman, M.; Hasan, M.M.; Famim, A.M.; Idrus, I.N.; Islam, M.T. Architecture of left-handed metamaterial absorber for absorbing electromagnetic hazards. *J. Optoelectron. Adv. Mater.* **2020**, *22*, 495–500.
10. Amer, A.A.G.; Sapuan, S.Z.; Nasimuddin, N.; Alphones, A.; Zinal, N.B. A comprehensive review of metasurface structures suitable for RF energy harvesting. *IEEE Access* **2020**, *8*, 76433–76452. [CrossRef]
11. Amer, A.A.G.; Sapuan, S.Z. Nasimuddin Multi-Band Metasurface Microwave Absorber Based on Square Split-Ring Resonator Structure. In *Proceedings of the 12th National Technical Seminar on Unmanned System Technology 2020*; Lecture Notes in Electrical Engineering; Springer: Singapore, 2022; Volume 770, pp. 373–382. [CrossRef]
12. Amer, A.A.G.; Sapuan, S.Z.; Nasimuddin, N.; Hassan, M.F. A Broadband Wide-Angle Metasurface Absorber for Energy Harvesting Applications. In Proceedings of the 2021 International Conference of Technology, Science and Administration (ICTSA), Taiz, Yemen, 22–24 March 2021. [CrossRef]
13. Bakır, M.; Karaaslan, M.; Unal, E.; Akgol, O.; Sabah, C. Microwave metamaterial absorber for sensing applications. *Opto-Electron. Rev.* **2017**, *25*, 318–325. [CrossRef]
14. Hajizadegan, M.; Ahmadi, V.; Sakhdari, M. Design and analysis of ultrafast and tunable all optical metamaterial switch enhanced by metal nanocomposite. *J. Light. Technol.* **2013**, *31*, 1877–1883. [CrossRef]
15. Xiong, H.; Hong, J.S.; Luo, C.M.; Zhong, L.L. An ultrathin and broadband metamaterial absorber using multi-layer structures. *J. Appl. Phys.* **2013**, *114*, 064109. [CrossRef]
16. Soheilifar, M.R.; Sadeghzadeh, R.A. Design, fabrication and characterization of stacked layers planar broadband metamaterial absorber at microwave frequency. *AEU-Int. J. Electron. Commun.* **2015**, *69*, 126–132. [CrossRef]
17. Tang, J.; Xiao, Z.; Xu, K.; Ma, X.; Wang, Z. Polarization-Controlled Metamaterial Absorber with Extremely Bandwidth and Wide Incidence Angle. *Plasmonics* **2016**, *11*, 1393–1399. [CrossRef]
18. Sheokand, H.; Ghosh, S.; Singh, G.; Saikia, M.; Srivastava, K.V.; Ramkumar, J.; Ramakrishna, S.A. Transparent broadband metamaterial absorber based on resistive films. *J. Appl. Phys.* **2017**, *122*, 105105. [CrossRef]
19. Cheng, Y.Z.; Wang, Y.; Nie, Y.; Gong, R.Z.; Xiong, X.; Wang, X. Design, fabrication and measurement of a broadband polarization-insensitive metamaterial absorber based on lumped elements. *J. Appl. Phys.* **2012**, *111*, 2010–2014. [CrossRef]

20. Yuan, W.; Cheng, Y. Low-frequency and broadband metamaterial absorber based on lumped elements: Design, characterization and experiment. *Appl. Phys. A Mater. Sci. Process.* **2014**, *117*, 1915–1921. [CrossRef]
21. Lim, D.; Lim, S. Ultrawideband Electromagnetic Absorber Using Sandwiched Broadband Metasurfaces. *IEEE Antennas Wirel. Propag. Lett.* **2019**, *18*, 1887–1891. [CrossRef]
22. Bağmancı, M.; Akgöl, O.; Özaktürk, M.; Karaaslan, M.; Ünal, E.; Bakır, M. Polarization independent broadband metamaterial absorber for microwave applications. *Int. J. RF Microw. Comput. Eng.* **2019**, *29*, e21630. [CrossRef]
23. Chen, K.; Luo, X.; Ding, G.; Zhao, J.; Feng, Y.; Jiang, T. Broadband microwave metamaterial absorber with lumped resistor loading. *EPJ Appl. Metamater.* **2019**, *6*, 1. [CrossRef]
24. Zhao, J.; Cheng, Y. Ultrabroadband Microwave Metamaterial Absorber Based on Electric SRR Loaded with Lumped Resistors. *J. Electron. Mater.* **2016**, *45*, 5033–5039. [CrossRef]
25. Banadaki, M.D.; Heidari, A.A.; Nakhkash, M. A Metamaterial Absorber with a New Compact Unit Cell. *IEEE Antennas Wirel. Propag. Lett.* **2018**, *17*, 205–208. [CrossRef]
26. Wang, Q.; Cheng, Y. Compact and low-frequency broadband microwave metamaterial absorber based on meander wire structure loaded resistors. *AEU-Int. J. Electron. Commun.* **2020**, *120*, 153198. [CrossRef]
27. Bakir, M.; Karaaslan, M.; Dincer, F.; Delihacioglu, K.; Sabah, C. Perfect metamaterial absorber-based energy harvesting and sensor applications in the industrial, scientific, and medical band. *Opt. Eng.* **2015**, *54*, 097102. [CrossRef]
28. Amer, A.A.G.; Sapuan, S.Z.; Nasimuddin, N. Efficient Metasurface Absorber for 2.4 GHz ISM-Band Applications. In Proceedings of the 2020 IEEE Student Conference on Research and Development (SCOReD), Batu Pahat, Malaysia, 27–29 September 2020; pp. 471–474. [CrossRef]
29. Amiri, M.; Tofigh, F.; Shariati, N.; Lipman, J.; Abolhasan, M. Miniature tri-wideband Sierpinski-Minkowski fractals metamaterial perfect absorber. *IET Microw. Antennas Propag.* **2019**, *13*, 991–996. [CrossRef]
30. Kalraiya, S.; Chaudhary, R.K.; Abdalla, M.A. Design and analysis of polarization independent conformal wideband metamaterial absorber using resistor loaded sector shaped resonators. *J. Appl. Phys.* **2019**, *125*, 134904. [CrossRef]
31. Kalraiya, S.; Chaudhary, R.K.; Abdalla, M.A. Resistor loaded wideband conformal metamaterial absorber for curved surfaces application. *AEU-Int. J. Electron. Commun.* **2022**, *143*, 154033. [CrossRef]
32. Jang, T.; Youn, H.; Shin, Y.J.; Guo, L.J. Transparent and Flexible Polarization-Independent Microwave Broadband Absorber. *ACS Photonics* **2014**, *1*, 279–284. [CrossRef]
33. Kong, X.; Xu, J.; Mo, J.J.; Liu, S. Broadband and conformal metamaterial absorber. *Front. Optoelectron.* **2017**, *10*, 124–131. [CrossRef]
34. Smith, D.R.; Vier, D.C.; Koschny, T.; Soukoulis, C.M. Electromagnetic parameter retrieval from inhomogeneous metamaterials. *Phys. Rev. E-Stat. Nonlinear Soft Matter Phys.* **2005**, *71*, 036617. [CrossRef]
35. Zuo, W.; Yang, Y.; He, X.; Zhan, D.; Zhang, Q. A miniaturized metamaterial absorber for ultrahigh-frequency RFID system. *IEEE Antennas Wirel. Propag. Lett.* **2017**, *16*, 329–332. [CrossRef]
36. Zuo, W.; Yang, Y.; He, X.; Mao, C.; Liu, T. An ultrawideband miniaturized metamaterial absorber in the ultrahigh-frequency range. *IEEE Antennas Wirel. Propag. Lett.* **2017**, *16*, 928–931. [CrossRef]

electronics

MDPI

Article

An Inverse Design Framework for Isotropic Metasurfaces Based on Representation Learning

Jian Zhang, Jin Yuan *, Chuanzhen Li and Bin Li

State Key Laboratory of Media Convergence and Communication, Communication University of China, Beijing 100024, China; zzjj1818@cuc.edu.cn (J.Z.); lichuanzhen@cuc.edu.cn (C.L.); libin08@cuc.edu.cn (B.L.)
* Correspondence: yuanjin@cuc.edu.cn

Abstract: A hybrid framework for solving the non-uniqueness problem in the inverse design of isomorphic metasurfaces is proposed. The framework consists of a representation learning (RL) module and a variational autoencoder-particle swarm optimization (VAE-PSO) algorithm module. The RL module is used to reduce the complex high-dimensional space into a low-dimensional space with obvious features, with the purpose of eliminating the many-to-one relationship between the original design space and response space. The VAE-PSO algorithm first encodes all meta-atoms into a continuous latent space through VAE and then applies PSO to search for an optimized latent vector whose corresponding metasurface fulfills the target response. This framework gives the solution paradigm of the ideal non-uniqueness situation, simplifies the complexity of the network, improves the running speed of the PSO algorithm, and obtains the global optimal solution with 94% accuracy on the test set.

Keywords: metasurfaces; representation learning; variational autoencoder; inverse design

Citation: Zhang, J.; Yuan, J.; Li, C.; Li, B. An Inverse Design Framework for Isotropic Metasurfaces Based on Representation Learning. *Electronics* **2022**, *11*, 1844. https://doi.org/10.3390/electronics11121844

Academic Editors: Naser Ojaroudi Parchin, Mohammad Ojaroudi and Raed A. Abd-Alhameed

Received: 13 May 2022
Accepted: 9 June 2022
Published: 10 June 2022

1. Introduction

Metasurfaces are synthetic composites composed of subwavelength structures arranged in different geometric shapes and distribution functions, which have been successfully applied in spectrum filtering, focusing, holographic imaging, polarization conversion, and other fields in recent years [1]. The subwavelength scatters that make up the metasurfaces are called meta-atoms, which resemble atoms or molecules of natural materials. When a beam of electromagnetic (EM) waves hits a metasurface, it is the strong resonances or spatial orientations of meta-atoms that cause the phase mutation, so metasurfaces have the ability to control the phase, amplitude, and polarization of reflected/transmitted waves in space [2].

Cui et al. proposed the concept of digital metasurface in 2014, in which the meta-atom pattern is discretized and encoded, greatly expanding the design space of metasurfaces [3]. Different coding sequences are proposed on this basis to achieve various EM responses, while the difficulty of inverse design has also increased dramatically with the number of codes for a digital metasurface increasing. Although some evolutionary algorithm (EA) has advantages in terms of fast random search without a problem domain [4], plenty of primary parameter settings of EAs have a dramatic impact on the evolution procedure and convergence result, which might lead to a fall in locally optimal solutions.

Fortunately, the development of deep learning, which is based on artificial neural networks to create computer reasoning by simulating human learning patterns in massive data, has brought new solutions [5]. Deep learning is a completely data-driven approach that mines implicit rules and relationships by learning from enough data sets, which has shown great advantages in computer vision, natural language processing, knowledge graph, and other fields. The basic idea is to design an algorithm to find rules between input data and output data according to a set of given data. The mined rules are stored in the deep learning model as the parameters given. If the rules between input and output remain unchanged, the model that has been trained can automatically and quickly predict its output as soon as another input is given. Since deep learning has its natural advantages in automatically mining undefined rules, we associate deep learning theory with metasurface inverse design to mine the relationship between the massive metasurface geometric parameters and response parameters. This could lead to a disruptive breakthrough in the field of metasurface design, skipping the physical interpretation and being faster and more efficient. Thus, there will be no professional theoretical requirements on designers so engineers are only required to pay attention to their practical demands instead of to the complicated design process and obscure physical theory [6]. The most difficult aspect of employing deep learning in the inverse design of metasurfaces is the non-uniqueness challenge [7]. This means that the expected EM response may correspond to multiple design patterns. Another challenge in using deep learning to design a complex meta-atom is the large size of the response and design spaces resulting in the need to train large neural networks [8], which means more parameters are required and networks are difficult to train. Adequate sampling points are required to achieve the desired EM response over a wide band, which typically results in thousands of data points in the response space. How to deal with the large design space and response space is of great importance.

In this article, we propose a new framework for meta-atom structure inverse design based on representation learning by addressing both the non-uniqueness issue and network-size issue. This framework consists of a representation learning (RL) module and a variational autoencoder-particle swarm optimization (VAE-PSO) algorithm module, which can achieve a high accuracy on the test data set. We use a sufficient number of data sets to train two different kinds of autoencoder, embedding high-dimensional space into low-dimensional space with a low loss rate, and the trained final network can output reduced design vectors (RDV) according to the target response. Then VAE-PSO algorithm is used to search for the optimal solution according to the RDV output by the RL module in the continuous space generated by VAE. Transforming the non-uniqueness issue into an optimal solution can greatly reduce the verification time and design cost. In contrast to earlier work, the most notable technique in this work is that the complex high-dimensional space is transformed into the characteristic low-dimensional space by the RL module, which gives the optimal solution to the non-uniqueness problem, greatly reduces the network complexity, and improves the running speed of the optimization algorithm enormously.

2. Theory and Design

As different metasurfaces can achieve the same EM response, when the framework inputs the target response, which metasurface will be returned is the non-unique problem. In practical application, though different metasurfaces can generally produce a similar response, it is almost impossible to produce an identical response at each of the sampling points. Therefore, in the global design space, there must be a metasurface that could realize the response closest to the target response, which means a framework is expected to find the optimal solution. The overall framework flow chart is shown in Figure 1. The task of the RL module is to translate the target EM response into RDV through the representation learning network. Any target EM response corresponds to a unique RDV, while one RDV might correspond to more than one metasurface. To solve this many-to-one problem, the VAE-PSO module generates latent design space, where the PSO algorithm is applied to search globally for the optimal latent vector. The VAE decoder is used to decode the latent vector into the optimal metasurface.

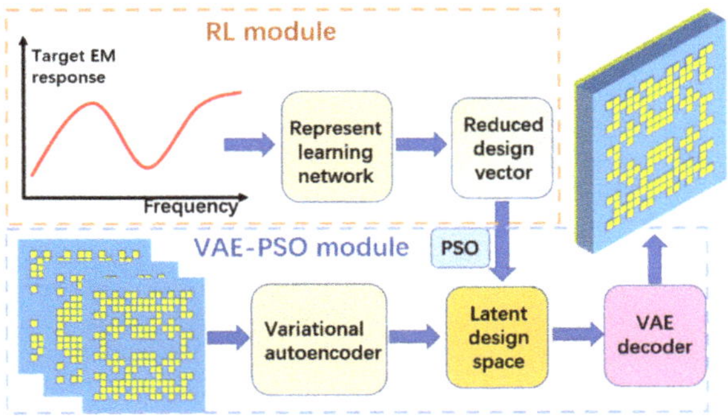

Figure 1. Flowchart of the hybrid framework.

2.1. Metasurface Design

Figure 2 illustrates the detailed structure of a meta-atom. The top pattern layer (Layer 1), with a side length L_1, is evenly divided into 16 × 16 discrete cells, where '1' means copper and '0' means vacuum in digital coding. Thus, the digital metasurface can be modeled as a 16 × 16 binary design matrix, which has $2^{8\times8}$ possible patterns. Each discrete copper piece is a square patch with a thickness of h_1 and a side length of L'_1. The second layer is a dielectric layer (Layer 2) with a dielectric constant of $\varepsilon_r = 2.65$, and a loss tangent of $\tan\delta = 0.001$. The back layer (Layer 3) is the backing copper sheet. The thickness and the side length of Layer 2 and Layer 3 are h_2, h_3, $L_{2,}$ and L_3, respectively. To reduce the influence of polarization, the object of our study is an isotropic metasurface composed of an 8 × 8 coding sequence as a subblock, which forms a 16 × 16 matrix through four-fold symmetry [9].

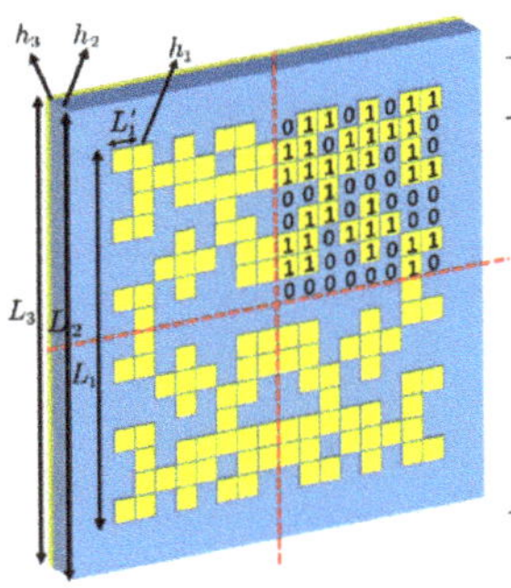

	Parameters	Values(mm)
Layer1	L_1	8.0
	h_1	0.017
	L'_1	0.5
Layer2	L_2	10.0
	h_2	2.0
Layer3	L_3	10.0
	h_3	0.017

Figure 2. Schematic illustration of a meta-atom structure.

In the simulation, boundary conditions and excitations are added to the metasurface model. The calculation principle of the software is solved by Maxwell's equations based on meshing, whose calculation time surges with the increase of model complexity. The design matrices of meta-atoms are selected as the input features and the S-parameter as the EM responses, in which data sets are randomly collected to train deep learning models. MATLAB to control CST STUDIO is used to generate digital metasurfaces, calculate S parameters, and save data sets automatically.

2.2. Representation Learning Module

The key to our framework is transforming design space and response space into low dimensional space by the RL module to solve the non-uniqueness problem and reduce network complexity [10]. First, the different spaces and their corresponding vectors need to be clearly defined: The original space includes the original design space (ODS) and original response space (ORS), and the corresponding vectors are the original design vector (ODV) and original response vector (ORV). The reduced space includes reduced design space (RDS) and reduced response space (RRS), and the corresponding vectors are reduced design vector (RDV) and reduced response vector (RRV). As multiple sets of meta-atoms can result in the same EM response, it is desired to map to the same vectors in the reduced space so that RDS and RRS can form a one-to-one mapping controlled by a nonlinear function, and this process is invertible. In this way, we transfer the many-to-one relationship between ODS and ORS into the many-to-one relationship between ODS and RDS, which could be solved by the VAE-PSO module. Figure 3 illustrates the mapping relationship between different spaces, with the red line representing one-to-one mapping and the blue line representing many-to-one mapping.

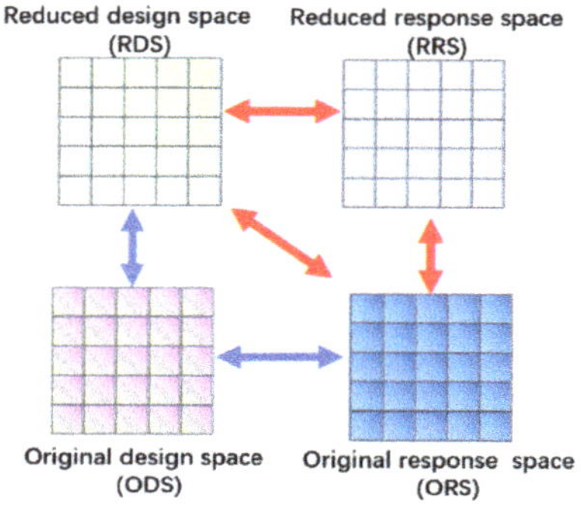

Figure 3. The relationship between different spaces. The red line represents one-to-one mapping and the blue line represents many-to-one mapping.

Representation learning is a machine learning technology that extracts effective features from raw data. Its essence is a dimensionality reduction technology, which can be realized by random forest, principal component analysis, autoencoder (AE), etc. In this article, the RL module uses AE, which is composed of an encoder and a decoder that can be used separately. The back propagation algorithm is used to continually train AE in order to minimize the cost function during the training process. Equations (1)–(3) illustrate features h, reconstruct data $\widetilde{x}$ and cost function L.

$$h = \sigma(W_{enc}x + b_{enc}) \tag{1}$$

$$\widetilde{x} = \sigma(W_{dec}h + b_{dec}) \tag{2}$$

$$L = \frac{1}{N}\sum_{i} \|x_i - \widetilde{x}_i\|_2^2 \tag{3}$$

where x represents the input data, W, b represents the transformation matrix and bias of the encoder and decoder, and N represents the amount of training data. σ is the active function. The dimension of h should be less than x in order to achieve feature dimension reduction.

Figure 4 illustrates the design process of AEs. The dimension of the ORS is reduced by training the autoencoder shown in Figure 4a. S11 between 0–20 GHz is selected as the EM response of this study, with a sampling frequency of 0.02 GHz and a length of 1000 dimensions. Subsequently, a pseudo-autoencoder (the input space is different from the output space) in Figure 4b is used to train the mapping between ODS and RDS, RDS and RRS together. The mapping of the former is many-to-one, and that of the latter is one-to-one. The one-to-one relationship between RDS and RRS is trained by a multi-layer perceptron (MLP). The decoder uses the "decoder1" trained in Figure 4a.

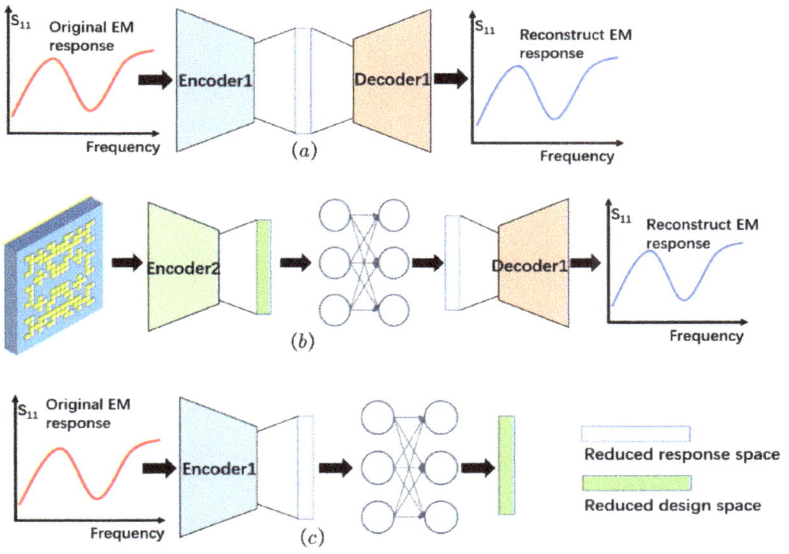

Figure 4. (**a**) Network 1, reduce the dimension of the ORS. (**b**) Network 2, used to forward prediction. (**c**) Network 3, RL module's ultimate network.

When the two reduced spaces are completed by a pseudo-autoencoder, an effective forward prediction model can be formed. It maps different design matrices that can achieve the same EM response to the same RDV, then maps them to the RRS one-to-one through MLP, and finally decodes ORV of 1000 dimensions. Subsequently, the inverse design model is built, and the inverse of the MLP should be found in Figure 4b. As demonstrated in Figure 4c, an inverse design network composed of the encoder1 in Figure 4a and the inverse structure of the MLP in Figure 4b is established. It can generate unique RDV by inputting a 1000-dimensional ORV.

The internal structures of encoders and decoders are various. A convolutional neural network (CNN) and a fully connected layer (FCL) are utilized in this work. Network 1 in Figure 4a is an autoencoder that reduces ORV of 1000 dimensions. If all FCL is adopted, the network scale and network parameters will be greatly increased, which is difficult to train. Thus, a hybrid network of one-dimensional convolution layer (Conv1D), pooling layer, and FCL are adopted in encoder1, and FCL is used in decoder1 in Figure 5a. Network 2 in Figure 4b is a pseudo-autoencoder, which transforms the design matrix into a reduced design vector and established a one-to-one mapping with the reduced response vector using MLP. The aim of network 2 is to train an encoder that can effectively compress an original design matrix to the RDV and train a MLP that connects two reduced spaces. Well-trained network 2 is a forward prediction network. CNN shows good feature extraction effect when processing 2D raster data [11], so our encoder2 uses a two-dimensional convolution layer (Conv2D) as the main network. The specific network structure is shown in Figure 5b. In this article, we collected 30,480 samples, among them, 21,366 were used for the training set, 6096 for the validation set, and 3018 for the test set. The training of these networks needs to strictly follow the order shown in Figure 5.

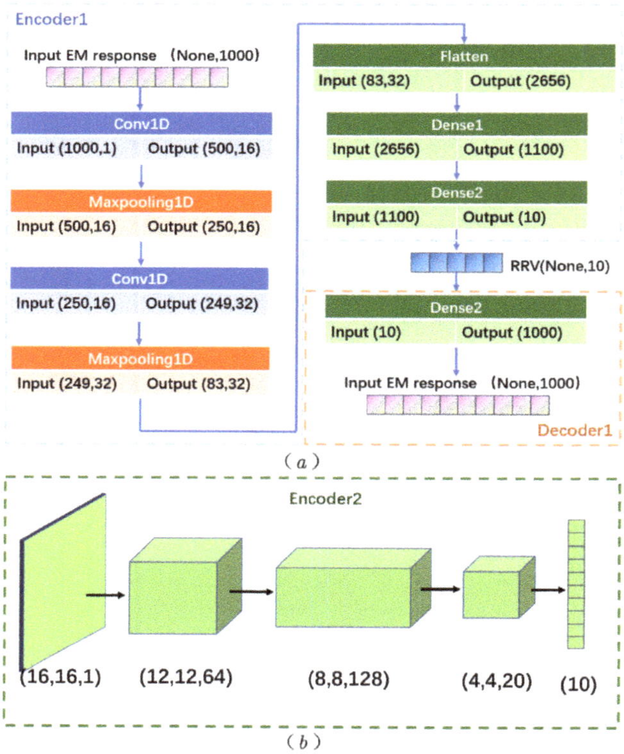

Figure 5. (**a**) The internal structure of encoder1, decoder1. (**b**) The internal structure of encoder2.

2.3. Variational Autoencoder-Particle Swarm Optimization Module

Although the final inverse design network can generate the required RDV, we still cannot find all the design matrices from the RDV due to the one-to-many relationship. Theoretically, a RDV may correspond to a multiple design matrix due to the non-uniqueness problem, and there is no suitable deep learning model that can carry out one-to-many mapping. In recent years, optimization algorithms have been widely used in the design of metasurfaces, but their running speed is slow due to the huge computational cost of original space [12]. Since the RL module has mapped the inverse design problem to a reduced space, a Particle Swarm Optimization (PSO) is applied in this low dimensional space with the purpose to improve the speed of the algorithm.

First, the global vector space of the 16×16 meta-atom space is generated to identify the target vector effectively and expeditiously in a global manner. A generative model can well complete this task [13], for which VAE is chosen to complete the generation of global vector space. A VAE can convert high-dimensional discrete data to a low-dimensional continuous space known as the latent space, where any sampling point can be encoded as a meaningful output. The VAE training process requires continuous learning of conditional probability distributions of inputs given latent variables based on the input data. When VAE is trained with the available binary matrix of 30,000 different isotropic metasurfaces, a continuous low-dimensional design space can be obtained and the reduced dimension is K [14]. Figure 6a illustrates an architecture of a VAE. Since the design matrix of the metasurface is a two-dimensional grid shape, the V-encoder and V-decoder of VAE are implemented by a convolutional neural network, which is illustrated in Figure 6b,c respectively.

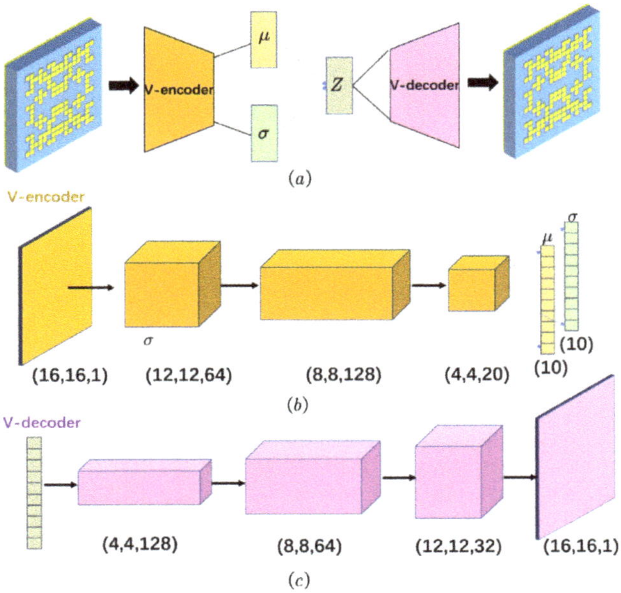

Figure 6. (**a**) The structure of variational autoencoder. (**b**) The internal structure of V-encoder. (**c**) The internal structure of V-decoder.

The encoder first converts the input binary matrix into mean vectors μ and standard deviation vectors σ. The latent vector Z is sampled from the Gaussian distribution $Z \sim N(\mu, \sigma^2)$ during the training process. After that, the decoder G reconstructs Z into the design matrix. Pattern topologies of meta-atom with comparable features are mapped to the same region of latent space in this process, and similar decoding meta-atoms are constantly modified by disrupting latent vectors. A well-trained decoder can recover any hidden vector sampled in the Z space into a binary matrix like the training set, which accomplishes the purpose of generating new samples. Equations (4)–(6) illustrate the loss function of a VAE, denoted by L_{VAE}.

$$L_{VAE} = L_{rec} + L_{KL} \tag{4}$$

$$L_{rec} = |x - \widetilde{x}|^2 \tag{5}$$

$$\begin{aligned} L_{KL} &= \sum_{k=1}^{K} KL[N(\mu_k, \sigma_k), N(0,1)] \\ &= \frac{1}{2}\sum_{k=1}^{K}\left(\sigma_k^2 + \mu_k^2 - \ln \sigma_k^2 - 1\right) \end{aligned} \tag{6}$$

where x, $\widetilde{x}$ represent input data and reconstruct data. KL refers to Kullback-Leibler divergence. After training VAE, a new network φ can be trained to find many-to-one mappings between continuous design space V and RDV D in Figure 7a. As shown in Figure 7b, the many-to-one relation can be converted into a multi-solution problem in functions. This can be expressed in Equation (7):

$$D = \varphi(Z) Z = \varphi^{-1}(D) \tag{7}$$

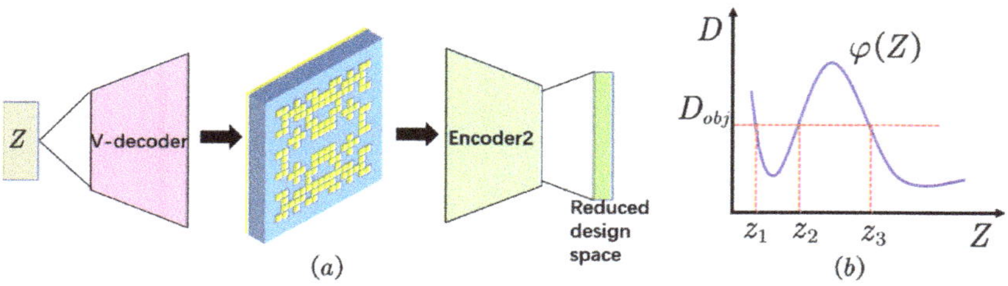

Figure 7. (**a**) A network for the VAE generated continuous latent vector into RDV. (**b**) The functional relationship between latent continuous vectors and RD.

Therefore, if the objective RDV and the function expression of the inverse network in Figure 5a are known, the non-uniqueness problem can be solved. Practically, it is impossible to express the function fitted with the neural network. Simultaneously, it is difficult for different meta-atoms to achieve the same response curve in every sample point. Therefore, it is necessary to introduce an optimization algorithm to find the optimal latent vectors with the smallest distance from the target RDV D_{tar}. Searching for the optimal solution to achieve the target response in the huge ODS will save a lot of verification work and greatly improve the inverse design efficiency of the metasurface. In this work, PSO was chosen as the optimization algorithm, which is implemented in Python using the PySwarms toolkit [15].

Figure 8 illustrates the optimization process. The continuous latent vector space is mapped to the RDS through the network φ. The fitness function is obtained by calculating the mapped vector D_φ and the target vector D_{tar} in RDS, which can be written in Equation (8).

$$F(z) = |\Delta D| = \left|D_{tar} - D_\varphi\right|^2 \tag{8}$$

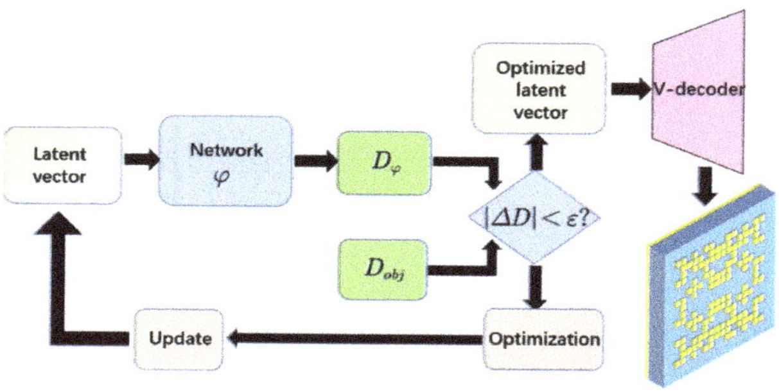

Figure 8. Flowchart of the VAE-PSO algorithm.

The fitness function is the objective function of optimization. The condition at the end of the optimization iteration is $|\Delta D| < \varepsilon$, where ε is an arbitrarily small value based on training data. After optimization, the optimal latent design vector is returned and the binary matrix can be generated by V-decoder. To evaluate the performance of our framework quantitatively [16], the accuracy of each target EM response is defined in Equation (9).

$$a = 1 - \frac{\int_{f_1}^{f_2} |R_{tar} - R_{gen}| df}{\int_{f_1}^{f_2} |R_{\text{tar}}| df} \tag{9}$$

where f_1 and f_2 are the frequency bounds of the input spectra, R_{tar} is the target EM response, R_{gen} is the generative response calculated by the generated design matrix from the VAE-PSO framework, which can be obtained directly using the trained network 2 in Figure 4b. The accuracy measures the matching degree between R_{tar} and R_{gen}. Applying Equation (9) to all test sets and then averaging them, the average accuracy of the entire framework is calculated. By transforming the non-uniqueness problem into a global optimal solution problem in lower-dimensional space, a high average accuracy of 94% on test sets can be achieved in our framework.

3. Design Result and Analysis

Initially, it is necessary to discuss whether network 1 learns the low-dimensional representation of ORS effectively. Network 1 is a typical AE for squeezing 1000 dimensional ORV down to 10 dimensions. Equation (10) illustrates the relative response error (RRE) to measure the quality of the autoencoder in the test set.

$$RRE = \left(\frac{\sqrt{\sum_{i=1}^{n} (r_i - g_i)^2}}{\sqrt{\sum_{i=1}^{n} r_i^2}} \right) \times 100\% \quad (10)$$

where r_i is the discretized value of ORV, g_i is the discretized value of the reconstruct EM response, and n is the number of discrete points of the spectrum. We input 1000 dimensions of the S11 curve into network 1 and the average RRE is 3.72% in the test set. The results of the two examples randomly selected are shown in Figure 9a,b. These results prove that a perfect consistency between the input response and the output response has been established and it is feasible to use RRV extracted by AE to represent the high-dimensional ORV.

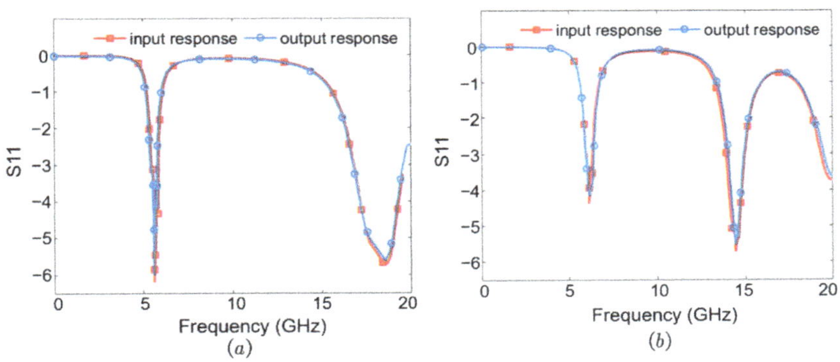

Figure 9. (**a**,**b**) Diagram of input response and output response of network 1.

Then, the performance of network 2 should be discussed. The essence of network 2 is a forward prediction network, which will input the design matrix of the metasurface and output the predicted ORV. Its main role in the overall framework is to quickly predict the response curve such as S11 of the inverse-designed metasurface instead of using a traditional physical simulator, so as to calculate the accuracy of this framework rapidly. Prediction accuracy is usually used to describe the performance of a forward prediction network and it can be expressed as 1-REE. Figure 10a,b illustrate the S11 curve respectively calculated by the physical simulator and the forward prediction network of two randomly selected metasurfaces. The average prediction accuracy of network 2 on the test set is up to 96%.

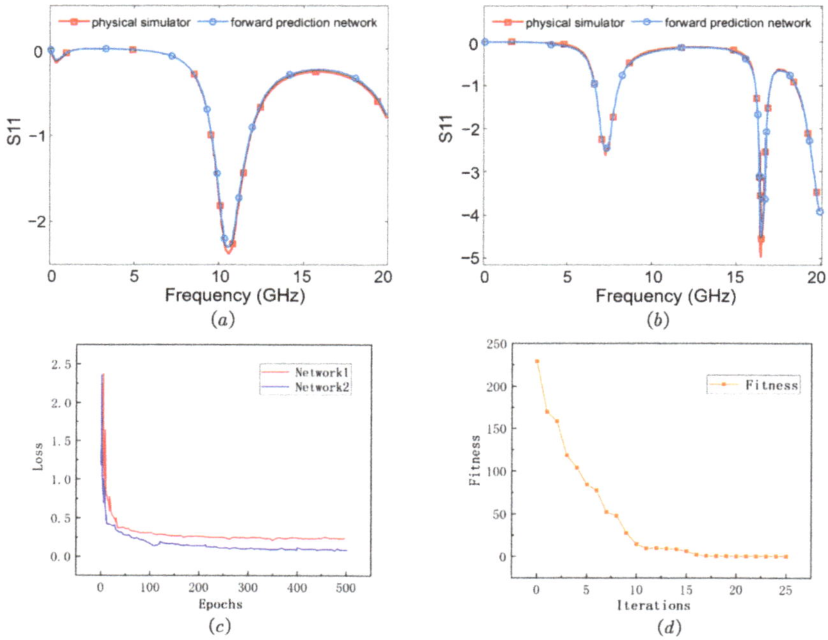

Figure 10. (**a**,**b**) The response curve calculated by the physical simulator and the prediction curve of the forward prediction network. (**c**) Validation loss of network 1 and network 2. (**d**) The process of VAE-PSO.

The training process of network 1 and network 2 is based on the back propagation algorithm, and the loss function value decreases rapidly with the number of epochs during the training process. Figure 10c shows the curve of the loss function for the two networks on the validation set. After about 25 epochs, the loss function values of the two networks can be reduced to less than 0.5, which achieved a good degree of training for an autoencoder network. The overall loss of network 1 is greater than that of network 2 because network 1 has high dimensions of input and output, which will be more difficult to train. To deal with the overfitting problem, L_2 regularization is necessary.

The VAE-PSO module realizes the application of the PSO algorithm in VAE generated latent space and the optimal latent vector is selected. The algorithm initializes 20 particles and sets the upper limit of iteration to 100. If the number of particles is too small, PSO will fall into the local optimal solution, while if the number of particles is too large, the running time will be greatly increased. It should be noted that the number of iterations should be set large enough to ensure the convergence of the PSO algorithm with the possibility of early stopping in case the fitness function does not change after about 15 iterations. Figure 10d illustrates how the fitness function changes during iteration. When iterating fifteen times, the fitness function reaches the minimum value and is almost unchanged in subsequent iterations, which proves the algorithm has converged.

Next, we introduce how to use the framework. Firstly, the target response is applied to network 3 to obtain the RDV, and then transmitted to the VAE-PSO algorithm framework and optimized iteratively. Finally, the design matrix that can produce the most similar EM response is output. As a verification, we select S11 curves as target responses from the test set to feed into the framework and get an average accuracy of up to 94%. Figure 11a illustrates a concrete example selected from the test set, with the target response in the blue line, and the generative response in the red line. For practical applications, perfectly ideal response curves made up of straight lines are usually input. We expect to design a metasurface capable of implementing a −2 dB S11 parameter at 12 GHz, so an ideal EM response without transition bands is fed into the framework and returns the most accurate design matrix in the global latent space in Figure 11b. An artificial neural network (ANN) can learn complex nonlinear input-output relationships through the training process and data adaptation, which are their main characteristics [17,18]. FCL is the most common type of ANN, which can directly train the corresponding rules between ORV and ODV. For performance comparison, the responses generated based on the FCL method, whose matching degree with the target response is far less than that of the framework, are also illustrated in Figure 11a,b. The phase response of the metasurface is also a significant property, which is illustrated in Figure 11c,d. Abrupt phase changes can be obtained in the planar metasurface structures over the subwavelength scale, which provides a new avenue to enable a variety of applications, including large-scale planar imaging, electromagnetic virtual shaping, and holographic display with a large field of view [19].

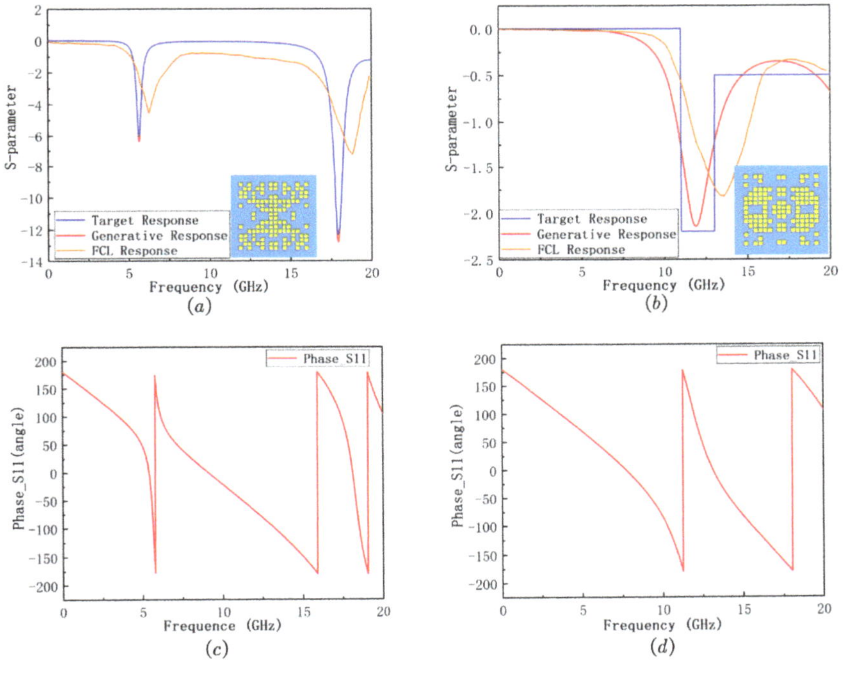

Figure 11. (**a**,**b**) Input the target EM response into the framework. (**c**,**d**) Phase curves corresponding to the metasurfaces of (**a**,**b**).

Finally, this framework is named "STARRY", and the performance difference among STARRY, the conventional design method, and the FCL design method is compared in Figure 12. Traditional design methods take genetic algorithms as an example. FCL design method directly trains the mapping of response curve to design matrix with full connection layer. The design time and area of deep learning methods are much smaller than that of the conventional design method. The STARRY framework consumes more time and space resources than FCL but shows the highest accuracy. At the same time, STARRY solves the problem of non-uniqueness and network complexity.

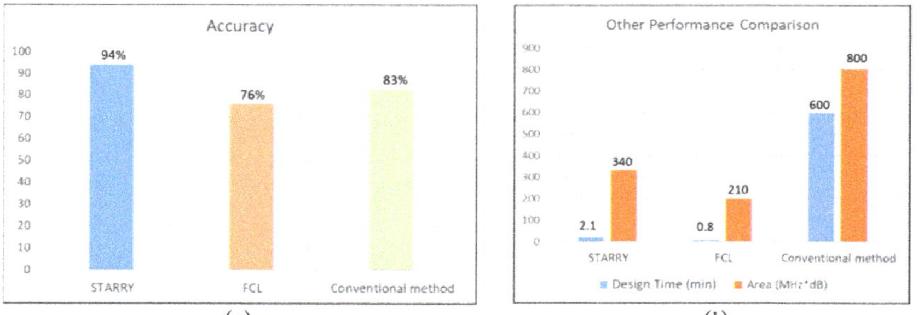

(*a*) (*b*)

Figure 12. (**a**) Accuracy comparison (**b**) Other performance comparison.

4. Conclusions

In conclusion, an isomorphic metasurfaces inverse design framework based on representation learning has been proposed and demonstrated. By using AEs of various architectures, the high-dimensional original space is mapped to the low-dimensional space with little loss, and the non-uniqueness problem is changed into a multi-solution problem in the low-dimensional space. The VAE-PSO algorithm is used to swiftly identify the best global search space value and deliver the binary matrix with the best response accuracy. This hybrid framework, which gives the optimal solution to the non-uniqueness problem and reduces the network complexity, achieves a high average accuracy of 94% on test sets, and it can give the design matrix in just a few seconds, which saves a lot of resources and time compared with the traditional inverse design methods.

Author Contributions: Conceptualization, J.Z.; methodology, J.Z. and J.Y.; software, J.Z.; validation, J.Z. and J.Y.; formal analysis, C.L.; investigation, B.L.; data curation, J.Z.; writing—original draft preparation, J.Z.; writing—review and editing, J.Y.; supervision, J.Y.; project administration, J.Y.; funding acquisition, J.Y. All authors have read and agreed to the published version of the manuscript.

Funding: National Natural Science Foundation of China (NSFC) (62105306). National undergraduate innovation and entrepreneurship training program of Communication University of China (202210033007). Fundamental Research Funds for the Central Universities (CUC220B004).

Acknowledgments: This work is supported by the National Natural Science Foundation of China (NSFC) (62105306), the National undergraduate innovation and entrepreneurship training program of Communication University of China (202210033007) and the Fundamental Research Funds for the Central Universities (CUC220B004). Jian Zhang and Jin Yuan thank the National Science Foundation for assistance in identifying collaborators for this work.

Conflicts of Interest: The authors declare no conflict of interest.

References

1. Boltasseva, A.; Atwater, H.A. Low-loss plasmonic metamaterials. *Science* **2011**, *331*, 290–291. [CrossRef] [PubMed]
2. Li, J.; Chen, S.; Yang, H.; Li, J.; Yu, P.; Cheng, H.; Gu, C.; Chen, H.-T.; Tian, J. Simultaneous control of light polarization and phase distributions using plasmonic metasurfaces. *Adv. Funct. Mater.* **2014**, *25*, 704–710. [CrossRef]
3. Cui, T.J.; Qi, M.Q.; Wan, X.; Zhao, J.; Cheng, Q. Coding metamaterials, digital metamaterials and programmable metamaterials. *Light. Sci. Appl.* **2014**, *3*, e218. [CrossRef]
4. Elsawy, M.M.R.; Lanteri, S.; Duvigneau, R.; Fan, J.A.; Genevet, P. Numerical optimization methods for metasurfaces. *Laser Photon.-Rev.* **2020**, *14*, 1900445. [CrossRef]
5. Hou, J.; Lin, H.; Xu, W.; Tian, Y.; Wang, Y.; Shi, X.; Deng, F.; Chen, L. Customized inverse design of metamaterial absorber based on target-driven deeplearning method. *IEEE Access* **2020**, *8*, 211849–211859. [CrossRef]
6. Qiu, T.; Shi, X.; Wang, J.; Li, Y.; Qu, S.; Cheng, Q.; Cui, T.; Sui, S. Deep learning: A rapid and efficient route to automatic metasurface design. *Adv. Sci.* **2019**, *6*, 1900128. [CrossRef] [PubMed]
7. Nadell, C.C.; Huang, B.; Malof, J.M.; Padilla, W.J. Deep learning for accelerated all-dielectric metasurface design. *Opt. Express* **2019**, *27*, 27523–27535. [CrossRef] [PubMed]

8. Li, J.; Li, Y.; Cen, Y.; Zhang, C.; Luo, T.; Yang, D. Applications of Neural Networks for Spectrum Prediction and Inverse Design in the Terahertz Band. *IEEE Photon. J.* **2020**, *12*, 5900509. [CrossRef]
9. Zhu, R.; Qiu, T.; Wang, J.; Sui, S.; Hao, C.; Liu, T.; Li, Y.; Feng, M.; Zhang, A.; Qiu, C.-W.; et al. Phase-to-pattern inverse design paradigm for fast realization of functional metasurfaces via transfer learning. *Nat. Commun.* **2021**, *12*, 2974. [CrossRef] [PubMed]
10. Kiarashinejad, Y.; Abdollahramezani, S.; Adibi, A. Deep learning approach based on dimensionality reduction for designing electromagnetic nanostructures. *NPJ Comput. Mater.* **2020**, *6*, 12. [CrossRef]
11. Sajedian, I.; Kim, J.; Rho, J. Finding the optical properties of plasmonic structures by image processing using a combination of convolutional neural networks and recurrent neural networks. *Microsyst. Nanoeng.* **2019**, *5*, 27. [CrossRef] [PubMed]
12. Diest, K. *Numerical Methods for Metamaterial Design*; Springer: Dordrecht, The Netherlands, 2013; Volume 127.
13. Liu, Z.; Zhu, D.; Rodrigues, S.P.; Lee, K.-T.; Cai, W. Generative model for the inverse design of metasurfaces. *Nano Lett.* **2018**, *18*, 6570–6576. [CrossRef] [PubMed]
14. Naseri, P.; Hum, S.V. A generative machine learning-based approach for inverse design of multilayer metasurfaces. *IEEE Trans. Antennas Propag.* **2021**, *69*, 5725–5739. [CrossRef]
15. Miranda, L.J.V. PySwarms: A research toolkit for particle swarm optimization in python. *J. Open Source Softw.* **2018**, *3*, 433. [CrossRef]
16. Ma, W.; Cheng, F.; Liu, Y. Deep-Learning-Enabled On-Demand design of chiral metamaterials. *ACS Nano* **2018**, *12*, 6326–6334. [CrossRef] [PubMed]
17. Abdolrazzaghi, M.; Hashemy, S.; Abdolali, A. Fast-forward solver for inhomogeneous media using machine learning methods: Artificial neural network, support vector machine and fuzzy logic. *Neural Comput. Appl.* **2016**, *29*, 1583–1591. [CrossRef]
18. Jamalian, M.; Safari, H.; Goodarzi, M. Permeability Prediction Using Artificial Neural Network and Least Square Support Vector Machine Methods. In Proceedings of the 80th EAGE Conference and Exhibition 2018, Copenhagen, Denmark, 11 June 2018.
19. Li, X.; Ma, X.; Luo, X. Principles and applications of metasurfaces with phase modulation. *Opto-Electron. Eng.* **2017**, *44*, 255–275.

electronics

Article

High-Performance Magnetoinductive Directional Filters

Artem Voronov *, Richard R. A. Syms and Oleksiy Sydoruk

Optical and Semiconductor Devices Group, EEE Department, Imperial College London, Exhibition Road, London SW7 2AZ, UK; r.syms@imperial.ac.uk (R.R.A.S.); o.sydoruk@imperial.ac.uk (O.S.)
* Correspondence: artem.voronov14@imperial.ac.uk

Abstract: Multiport magnetoinductive (MI) devices with directional filter properties are presented. Design equations are developed and solved using wave analysis and dispersion theory, and it is shown that high-performance directional filters can be realised for use both in MI systems with complex, frequency-dependent impedance and in conventional systems with real impedance. Wave analysis is used to reduce the complexity of circuit equations. High-performance MI structures combining directional and infinite rejection filtering are demonstrated, as well as multiple-passband high-rejection filtering. A new method for improving filtering performance through multipath loss compensation is described. Methods for constructing tuneable devices using toroidal ferrite-cored transformers are proposed and demonstrated, and experimental results for tuneable MI directional filters are shown to agree with theoretical models. Limitations are explored, and power handling sufficient for HF RFID applications is demonstrated, despite the use of ferrite materials.

Keywords: directional filter; infinite rejection; magnetoinductive waveguide; metamaterial

Citation: Voronov, A.; Syms, R.R.A.; Sydoruk, O. High-Performance Magnetoinductive Directional Filters. *Electronics* **2022**, *11*, 845. https://doi.org/10.3390/electronics11060845

Academic Editors: Naser Ojaroudi Parchin, Mohammad Ojaroudi and Raed A. Abd-Alhameed

Received: 28 January 2022
Accepted: 4 March 2022
Published: 8 March 2022

1. Introduction

Directional filters (DFs) are four-port directional couplers with frequency filtering capability [1–4]. They can remove band-limited noise or unwanted signals at specific frequencies or be arranged in cascade as multiplexers for frequency-division multiplexing [5–7]. Lumped element and waveguide formats have been developed for frequencies ranging from VHF to millimetre wave [8–15]. Two-port filters with high attenuation have also been developed and use multipath cancellation, reflection, or absorption to achieve infinite rejection [16–23]. Applications include transmit/receive modules, multiplexers in ultra-wide-band antennas, and superheterodyne receivers [24–28]. Apart from rare examples [29], development has involved systems with real impedance. However, work on RF and microwave metamaterials showed that (in addition to other novel properties) periodic arrays of coupled metallic resonators allow the propagation of lattice waves including electroinductive (EI) [30–33] and magnetoinductive (MI) [34,35] waves. The latter have attracted considerably more attention for applications that depend on magnetic rather than electric fields. These include near-field communication in weakly conductive media [36–38], inductive power transfer [39–41], and magnetic field sensing [42–46]. MI waveguides consist of arrays of magnetically coupled LC resonators. They are band-limited and dispersive and have complex frequency-dependent impedance. Although quasi-optical devices, such as matching networks, mirrors, resonators splitters, and couplers, can be synthesised [47–49], MI systems are still embryonic, and it is difficult to integrate other functionality, such as filtering.

Here we show that a pair of coupled right- and left-handed MI waveguides can form a directional filter and investigate its potential for signal blocking in MI-based high-frequency radio frequency identification (HF RFID), where MI antennas may also find application [46]. The filters are designed to effectively operate at high RF power, since operation is based on complementary outputs rather than absorption, and to ensure effective blocking, frequency selectivity is tuneable. Furthermore, a new method for realising four-port directional filters with tuneable infinite-rejection responses is presented. Wave theory is developed to explain the principle of multipath loss compensation, and high attenuation is confirmed by experiment. Designs are extended to allow multiple stopbands with improved rejection. Three-port directional filters are also shown to be feasible and can be made using a section of MI waveguide with the addition of as few as three resonators. In each case, wave theory is used to reduce the complexity of the circuit equations. Simple multiplexers can be implemented by cascading directional filters. Together, these features can introduce additional functionality and complexity in magnetoinductive systems, since the filter components can be integrated directly with MI waveguides. Additionally, such devices can be successfully used in conventional systems.

Essential background is first reviewed in Sections 2.1 and 2.2. An analytic model for MI directional filters is presented in Section 2.3, the circuit equations are solved using discrete travelling wave analysis, and methods for tuning are proposed. Conditions for infinite rejection are identified in Section 2.4, and the theory is extended to multiple passbands. Methods for constructing tuneable devices with high rejection and multiple passbands are described. Theoretical predictions are shown to be in good agreement with experimental results in the HF band in Section 3. Performance parameters including power handling and frequency scaling are also considered, with conclusions in Section 4.

2. Materials and Methods

2.1. Magnetoinductive Waveguides

Figure 1a is a schematic of a magnetoinductive waveguide, which consists of magnetically coupled resonators with inductance L, capacitance C, resistance R, and nearest-neighbour mutual inductance M. Although non-nearest-neighbour coupling is a common problem, we assume that steps have been taken to avoid it. Assuming that the loop current in resonator n is I_n, the circuit equations relating the currents in adjacent elements are:

$$\left(j\omega L + \frac{1}{j\omega C} + R\right) I_n + j\omega M(I_{n-1} + I_{n+1}) = 0. \quad (1)$$

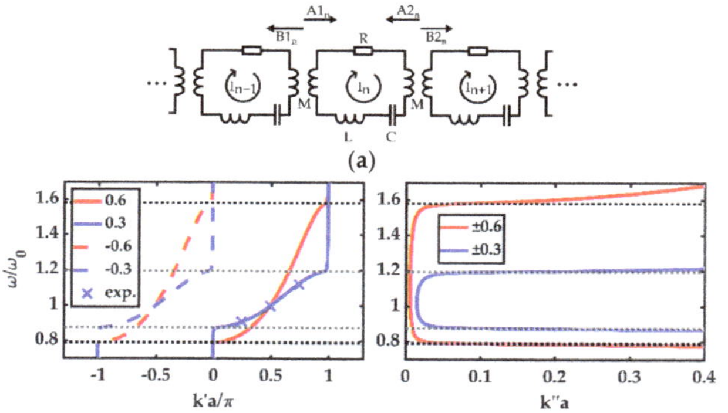

(b)

Figure 1. Magnetoinductive (MI) waveguide: (**a**) equivalent circuit; (**b**) frequency dependence of $k'a$ and $k''a$ for $\kappa = \{\pm 0.6, \pm 0.3\}$ and $Q = 200$. Crosses show measured points for $\kappa = 0.3$.

For convenience, we define the self-impedance $Z = j\omega L + \frac{1}{j\omega C} + R$, mutual impedance $Z_m = j\omega M$, and magnetic coupling coefficient $\kappa = \frac{2M}{L}$. Assumption of a travelling current wave in the form $I_n = I_0 e^{-jkna}$, where $k = k' - jk''$ is the complex propagation constant and a is the element spacing, yields the well-known dispersion relation [34]:

$$1 - \frac{\omega_0^2}{\omega^2} - \frac{j}{Q}\frac{\omega_0}{\omega} + \kappa\cos(ka) = 0. \quad (2)$$

Here $\omega_0 = \frac{1}{\sqrt{LC}}$ is the angular resonant frequency and $Q = \frac{\omega_0 L}{R}$ is the quality factor. For lossless waveguides, the Q-factor is infinite, and the left-hand side of Equation (2) is real with $k'' = 0$. In real systems, Q must be finite. If losses are low ($k'' \ll k'$), dispersion and loss may be approximated using:

$$1 - \frac{\omega_0^2}{\omega^2} + \kappa \cos(k'a) = 0 \quad (3)$$

$$k''a = \frac{\omega_0}{\kappa Q \omega \sin(k'a)}. \quad (4)$$

Equation (3) is the dispersion relation for lossless guides. It implies that MI waveguides are band limited, with propagation in the range $\frac{1}{\sqrt{1+|\kappa|}} \leq \frac{\omega}{\omega_0} < \frac{1}{\sqrt{1-|\kappa|}}$. Guides with positive M are right-handed and have positive phase velocity, while negative M yields left-handed guides with negative phase velocity. At the resonant frequency, $k'a = \frac{\pi}{2}$ when $M > 0$ and $-\frac{\pi}{2}$ when $M < 0$. Equation (4) implies that losses are inversely proportional to $|\kappa Q|$, are lowest at the resonant frequency, and rise rapidly at the band edges. Figure 1b shows the frequency dependence of $k'a$ and $k''a$ for $Q = 200$ and two different values of $|\kappa|$, which clearly show the band-limited nature of propagation. Crosses show three experimentally measured values along the dispersion relation for $\kappa = 0.3$ and are clearly in agreement with Equation (3).

2.2. Impedance and Power Flow

It is simple to show that the characteristic impedance of an MI waveguide is [47]:

$$Z_0 = j\omega M e^{-jka}. \quad (5)$$

If the waveguide is lossless, the characteristic impedance Z_{0M} at resonance is purely real, with $Z_{0M} = \omega_0 M$. In general, the impedance is complex valued and frequency dependent, which implies that care must be taken when matching to real-valued loads. An important consequence is that conventional scattering parameters cannot be used; instead, the formulation of 'power waves', originally developed by Kurokawa [50] and recently adapted to MI waveguides [51], is required. For MI lines extending to the left, type 1 power waves $A1$ and $B1$ are used at element n, as shown in Figure 1a, defined as:

$$A1 = \frac{-Z_m I_{n-1} + Z_0 I_n}{2\sqrt{Re(Z_0)}} \qquad B1 = \frac{-Z_m I_{n-1} - Z_0^* I_n}{2\sqrt{Re(Z_0)}} \quad (6)$$

Here Z_0^* represents the complex conjugate of Z_0, and $Re(Z_0)$ its real part. For MI lines extending to the right, type 2 power waves $A2$ and $B2$ must be used, defined as:

$$A2 = \frac{Z_m I_{n+1} - Z_0 I_n}{2\sqrt{Re(Z_0)}}, \qquad B2 = \frac{Z_m I_{n+1} + Z_0^* I_n}{2\sqrt{Re(Z_0)}}. \quad (7)$$

The squared moduli of these coefficients provide the correct definition of power.

2.3. Magnetoinductive Directional Filters

A directional filter has four ports connected to lines of characteristic impedance Z_0. A voltage source connected to port 1 then gives rise to incident and reflected waves $A1_1$ and $B1_1$ and transmitted waves $B2_2, B2_3, B1_4$ from ports 2, 3, and 4. Port 2 has the response of a band-reject filter, and Port 4 that of a bandpass filter. In a matched and lossless device, S_{21} and S_{41} are complementary, $S_{31} = 0$ for all ω, and $S_{11} = 0$ at resonance. Equivalent properties apply no matter which port is used as input. We now develop the analysis for MI directional filters and consider their frequency response, bandwidth, and tuneability.

Figure 2a shows the equivalent circuit of a filter based on a pair of infinite horizontal MI waveguides consisting of identical resonators, but with opposite coupling polarity so that the lower guide is right-handed (blue) and the upper guide left-handed (red). The mutual inductances are equal in magnitude, with $M > 0$, so the coupling coefficients are $\kappa_f = \frac{2M}{L}$ and $\kappa_b = -\frac{2M}{L}$. The waveguides are connected vertically via two resonators, which couple the guides via mutual inductances M_C and coupling coefficients $\kappa_c = \frac{2M_c}{L}$, such that $\kappa_c < \kappa_f$. The network has four ports, 1–4, formed by sections of an MI waveguide. The lower line supports currents I_n, and the top line supports J_n, while the two coupling resonators have currents K and P. Note that guide terminations can also be represented by lumped impedances, as shown in Figure 2b. There are three main possibilities: an infinite MI waveguide, represented by Z_0 (1); a resistive termination, represented by Z_{0M} (2); and a broadband termination to Z_{0M}, achieved by halving the inductance and doubling the capacitance in the resonant element (3). Termination 2 provides an impedance match at ω_0, while termination 3 provides matching at two frequencies, ω_0 and $\frac{\omega_0}{\sqrt{1-\kappa^2}}$, with low reflection in between [49].

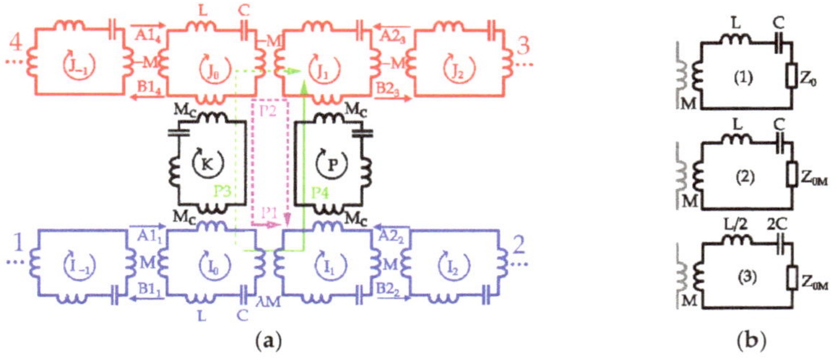

Figure 2. (**a**) MI directional filter equivalent circuit; (**b**) lumped element terminations: (1) magnetoinductive, (2) resistive, and (3) broadband resistive.

For now, we assume type 1 terminations and that an MI wave is incident on port 1 from the left. Experience suggests that this wave will be partially reflected at $n = 0$ due to the discontinuity caused by the coupling resonator, and that the remaining power, less any loss, will appear as MI waves from ports 2 to 4. We write expressions for the currents I_n and J_n in terms of incident, reflected, and transmitted MI waves as:

$$\begin{aligned} I_n &= Ie^{-jk_f na} + Re^{jk_f na},\ n \leq 0 \\ I_n &= T_2 e^{-jk_f na},\ n \geq 1 \\ J_n &= T_3 e^{-jk_b na},\ n \geq 1 \\ J_n &= T_4 e^{jk_b na},\ n \leq 0. \end{aligned} \tag{8}$$

Here I and R are the amplitudes of the incident and reflected waves at port 1; T_2, T_3, T_4 are the wave amplitudes from ports 2, 3, and 4; $k_f = k_f' - jk_f''$ is the propagation constant of the forward waveguide; and $k_b = k_b' - jk_b''$ is the propagation constant of the backward one. From Figure 1b and Equation (2), we can assume:

$$k_b a = k_f a - \pi. \tag{9}$$

Combining Equation (1) with circuit equations for the currents K and P in the coupling resonators, a set of coupled equations may be obtained at the six central resonators:

$$\begin{aligned} ZI_0 + Z_m(I_{-1} + I_1) + Z_{mc}K = 0, &\quad ZI_1 + Z_m(I_0 + I_2) + Z_{mc}P = 0 \\ ZK + Z_{mc}(I_0 + J_0) = 0, &\quad ZP + Z_{mc}(I_1 + J_1) = 0 \\ ZJ_0 - Z_m(J_{-1} + J_1) + Z_{mc}K = 0, &\quad ZJ_1 - Z_m(J_0 + J_2) + Z_{mc}P = 0. \end{aligned} \tag{10}$$

Substituting the assumed solutions and using the dispersion equation, K and P may be eliminated, allowing the reflection and transmission coefficients to be calculated:

$$\begin{aligned} \frac{R}{I} &= \frac{-\left(e^{j2k_f a}+1\right)\mu^2}{1+2\mu^2 e^{j2k_f a}-e^{j4k_f a}}, \quad \frac{T_4}{I} = \frac{\left(1-e^{j2k_f a}\right)\mu^2}{1+2\mu^2 e^{j2k_f a}-e^{j4k_f a}} \\ \frac{T_2}{I} &= \frac{1-e^{j4k_f a}}{1+2\mu^2 e^{j2k_f a}-e^{j4k_f a}}, \quad \frac{T_3}{I} = 0. \end{aligned} \tag{11}$$

Here $\mu = \frac{M_c}{M}$ is the mutual inductance ratio. Note that the coefficients above are complex valued and frequency dependent. They are expressed only in terms of μ and k_f, which may be found from Equation (2). However, care must be taken to ensure that $k_f'' \geq 0$ and $0 \leq k_f' \leq \pi$. The results imply that no signal leaves port 2 at resonance, provided the guides are lossless. Similarly, the transmission from port 4 is maximum at resonance, while the reflectance from port 1 is zero. Port 3 is isolated at all frequencies, both within and outside the MI passbands. This behaviour is inherent in the way signals are combined. The phase shift at resonance along path P1 in Figure 2a is $\frac{\pi}{2}$, while the phase shift along path P2 is $\frac{3\pi}{2}$. If the device is lossless, the signal from port 2 is then zeroed by cancellation. If Q-factors are finite, losses are different along P1 and P2 and cancellation is partial. Two signal paths, P3 and P4, to port 3 are in antiphase for all frequencies.

Although the results above provide valid amplitude reflection and transmission coefficients, they offer no insight on power flow in devices that are either lossy or operating out of band. We therefore define three sets of discrete power wave pairs, $\{A1_1, B1_1\}$, $\{A2_2, B2_2\}$, and $\{A1_4, B1_4\}$, for ports 1, 2, and 4, as shown in Figure 2a, as follows:

$$\begin{aligned} A1_1 &= \frac{-Z_m I_{-1}+Z_{0f} I_0}{2\sqrt{Re(Z_{0f})}}, \quad A2_2 = \frac{Z_m I_2 - Z_{0f} I_1}{2\sqrt{Re(Z_{0f})}}, \quad A1_4 = \frac{Z_m I_{-1}+Z_{0b} I_0}{2\sqrt{Re(Z_{0b})}}, \\ B1_1 &= \frac{-Z_m I_{-1}-Z_{0f}^* I_0}{2\sqrt{Re(Z_{0f})}}, \quad B2_2 = \frac{Z_m I_2 + Z_{0f}^* I_1}{2\sqrt{Re(Z_{0f})}}, \quad B1_4 = \frac{Z_m I_{-1}-Z_{0b}^* I_0}{2\sqrt{Re(Z_{0b})}}. \end{aligned} \tag{12}$$

Here $Z_{0f} = j\omega M e^{-jk_f a}$ and $Z_{0b} = -j\omega M e^{-jk_b a}$ are the characteristic impedances of the forward and backward guides. In fact, $Z_{0f} = Z_{0b}$ whenever Equation (9) applies. Power wave scattering parameters with respect to port 1 may then be found as:

$$S_{11} = \frac{B1_1}{A1_1}, \; S_{21} = \frac{B2_2}{A1_1}, \; S_{41} = \frac{B1_4}{A1_1}. \tag{13}$$

Using the transmission and reflection coefficients, we may obtain analytic expressions for the scattering parameters of a directional filter with MI line terminations (1) as:

$$\begin{aligned} S_{11} &= \left(\left(1-e^{2k_f'' a}\right)\left\{e^{j2k_f' a}+e^{2(k_f''+j2k_f')a}\right\}+\mu^2\left(1+2e^{2(k_f''+jk_f')a}-e^{j2k_f' a}\right)\right)/d \\ S_{21} &= \left(e^{(k_f''+jk_f')a}-e^{(k_f''+j3k_f')a}+e^{-(k_f''+jk_f')a}-e^{(-k_f''+jk_f')a}\right)/d \\ S_{41} &= \mu^2\left(e^{j2k_f' a}-1\right)/d, \\ d &= 1+2\mu^2 e^{2(k_f''+jk_f')a}-e^{4(k_f''+jk_f')a}. \end{aligned} \tag{14}$$

Moduli squared of Equations (11) and (14) are equal whenever $k_f'' = k_b'' = 0$.

2.3.1. Frequency Response

To illustrate the above, Figure 3a shows the frequency dependence of magnitude S-parameters for different terminations, with $L = 1\ \mu H$, $\kappa = 0.6$, and $\kappa_c = 0.3$ (so that $\mu = 1/2$). Results for lossless and lossy elements with magnetoinductive terminations are obtained using Equation (14), and results for lossy elements with narrowband resistive terminations using numerical solution of circuit equations.

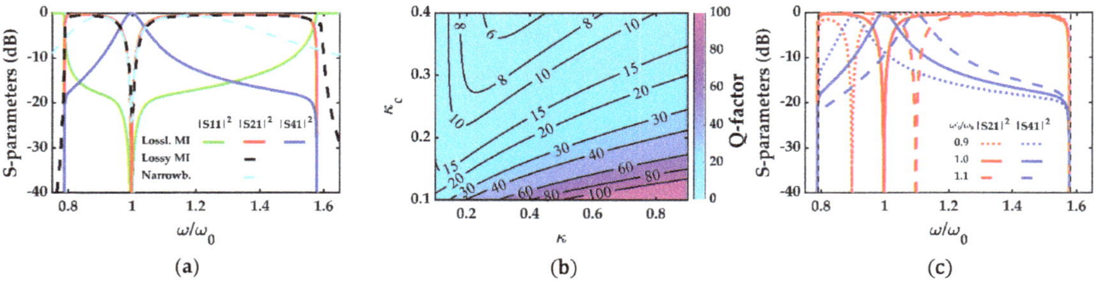

Figure 3. MI directional filter performance for $\kappa = 0.6$ and $\kappa_c = 0.3$: (**a**) scattering parameters for lossless and lossy ($Q = 200$) resonators with MI terminations (1) and real narrowband terminations (2); (**b**) variation of Q_{bp} with κ and κ_c for a lossless filter with MI terminations (1); (**c**) scattering parameters for lossless resonators with MI terminations (1) for $\omega_0'/\omega_0 = 0.9$, 1.0, and 1.1.

In each case, frequencies near ω_0 are diverted into port 4, while the remaining power is sent to port 2. Port 4 is therefore bandpass, while port 2 is a bandstop. The MI passband is as estimated from previous arguments. The isolation of port 3 is unaffected by loss; however, power emerging from port 4 is reduced and port 2 rejection diminished. Since losses are inversely proportional to κQ and $\kappa_c Q$, high coupling coefficients and Q-factors are required to minimise loss. The effect of a narrowband resistive termination is to reduce the bandwidth. However, this can be partly restored by using a broadband termination.

2.3.2. Bandwidth

An analytic expression for the 3-dB bandwidth of $|S_{41}|^2$ may be obtained in the lossless case. The half-power frequencies can be found as follows:

$$|S_{41}|^2_{k''=0} = \frac{\mu^4 e^{j2k_f' a}\left(e^{j2k_f' a} - 1\right)^2}{1 - e^{j4k_f' a}\left(4\mu^4 + 2 - e^{j4k_f' a}\right)} = \frac{1}{2}. \tag{15}$$

Within the passband, there are two solutions for k_f'. If these are expressed in terms of κ, μ, and ω_0, the half-power bandwidth B is given by:

$$B = \omega_0\left(\sqrt{\frac{1}{1-\kappa p}} - \sqrt{\frac{1}{1+\kappa p}}\right), \quad p = \frac{\sqrt{\mu^4 - \sqrt{\mu^8 + 4} + 2}}{2}. \tag{16}$$

Thus, the width of the bandpass channel increases with μ and ω_0. It is simple to calculate the Q-factor of the bandpass response, as $Q_{bp} = \frac{\omega_0}{B}$. Figure 3b shows a contour map of the variation of Q_{bp} with κ and κ_c thus obtained. High-filter Q-factors require high values of κ and low values of κ_c; however, the achievable Q-factor is limited since the use of real resonators with finite Q-factors will result in high losses if κ_c is too low.

2.3.3. Frequency Tuning

So far, the filter centre frequency has been determined by ω_0, which also determines the operation of the MI waveguides. It would clearly be beneficial to introduce tuning within the MI passband. Figure 3c shows the result of tuning the resonance of the coupling resonators to different values, namely, (a) $\omega_0' = 0.9\omega_0$, (b) $\omega_0' = 0.95\omega_0$, and (c) $\omega_0' = 1.1\omega_0$, assuming lossless resonators and lossless MI terminations. Although the best performance is obtained when $\omega_0' = \omega_0$, tuning around ω_0 is feasible for the variations in ω_0' shown.

From the above, we conclude that the network shown in Figure 2a with lossless terminations possesses all the properties of a directional filter except that it has null reflectivity only within the MI passband. S_{11} is zero at the resonant frequency and very small in its vicinity. Lossy systems and systems with narrowband terminations have similar properties, the main difference being a reduction in the magnetoinductive bandwidth.

2.4. Advanced Filters

In this section, we consider the properties of more advanced filters, which offer either infinite rejection or multiple bandstop frequencies.

2.4.1. Infinite Rejection

In the absence of loss, complete rejection occurs at ω_0 at port 2 when port 1 is used as an input. However, for finite Q-factors, the two signal paths from port 1 to port 2 in Figure 2a incur unequal losses, and some leakage into port 2 then occurs even at resonance. To equalise signal amplitudes along the two paths, the magnetic coupling strength between resonators 0 and 1 in the forward waveguide may be adjusted. We define a new positive coefficient, $\kappa_f' = \lambda\kappa_f$, between these elements and reconsider the port reflection and transmission coefficients. The circuit equations must be modified to:

$$\begin{gathered} ZI_0 + Z_m(I_{-1} + \lambda I_1) + Z_{mc}K = 0,\ ZI_1 + Z_m(\lambda I_0 + I_2) + Z_{mc}P = 0 \\ ZK + Z_{mc}(I_0 + J_0) = 0,\ ZP + Z_{mc}(I_1 + J_1) = 0 \\ ZJ_0 - Z_m(J_{-1} + J_1) + Z_{mc}K = 0,\ ZJ_1 - Z_m(J_0 + J_2) + Z_{mc}P = 0. \end{gathered} \tag{17}$$

As before, we may substitute the assumed solutions and use the dispersion relation to obtain modified reflection and transmission coefficients. The transmission from port 2 may then be expressed as a rational polynomial in λ as follows:

$$\begin{gathered} \frac{T_2}{I} = \frac{p_0 + \lambda p_1}{q_0 + 2\mu^4 e^{j2k_f a}\lambda + q_2\lambda^2} \\ p_0 = \mu^4 e^{j2k_f a}\left(1 - e^{j2k_f a}\right) \\ p_1 = \left(e^{j4k_f a} - 1\right)^2 + \mu^2 e^{j2k_f a}\left(2 - 2e^{j4k_f a} + \mu^2\left(e^{j2k_f a} - 1\right)\right) \\ q_0 = e^{j2k_f a}\left(e^{j6k_f a} + e^{j4k_f a} - e^{j2k_f a} - 1\right) + \mu^2 e^{j2k_f a}\left(2 - 2e^{j2k_f a} - 4e^{j4k_f a} + 4\mu^2 e^{j2k_f a} - \mu^2\right) \\ q_2 = 1 + e^{j2k_f a} - e^{j4k_f a} - e^{j6k_f a} + \mu^2 e^{j2k_f a}\left(2 + 2e^{j2k_f a} - \mu^2\right). \end{gathered} \tag{18}$$

For infinite rejection at midband, we require the numerator of T_2/I to equal zero:

$$\lambda = -\frac{p_0}{p_1}. \tag{19}$$

At resonance, the propagation constant can be rewritten in terms of κ and Q alone:

$$k_f a\Big|_{\omega=\omega_0} = \frac{\pi}{2} + jln\left(\sqrt{1 + \frac{1}{\kappa^2 Q^2}} - \frac{1}{\kappa Q}\right). \tag{20}$$

Substituting into Equation (19), we obtain the condition for infinite rejection as:

$$\lambda = \frac{\mu^4\rho^4}{\mu^4\rho^4 - 2\mu^2\rho^4 + 2\mu^2\rho^2 + \rho^6 - \rho^4 - \rho^2 + 1},\quad \rho = \sqrt{1 + \frac{1}{\kappa^2 Q^2}} - \frac{1}{\kappa Q}. \tag{21}$$

Figure 4a shows the variation of λ with μ for a range of Q-factors. Lower values of λ are required for higher losses, as well as for lower values of μ. In each case, $\lambda \to 1$ as μ increases, with sharper increases at lower losses. When λ is set as above, port 2 is isolated at resonance. However, there is an impact on filter performance; namely, port 3 is no longer fully isolated, the passband of port 2 reduces, and reflectance at port 1 is improved.

Figure 4b shows the frequency dependence of scattering parameters for filters with lossy MI (1) and resistive narrowband (2) terminations for $\kappa = 0.6$, $\kappa_c = 0.3$, $Q = 200$, and $\lambda = 0.8746$ from Equation (21). Both cases demonstrate excellent directional filtering with reflectance mostly <-30 dB within the passband, as well as infinite rejection at resonance. If isolation is important, λ may be adjusted to balance rejection with leakage into port 3. The mutual inductance discontinuity generated by tuning λ implies that the complementary waveguide pair is no longer symmetrical, meaning the ports cannot be used interchangeably. Only the two ports immediately adjacent to the λ-tuned resonator pair would yield an infinite-rejection response. The remaining two ports would suffer from reduced performance if used as inputs.

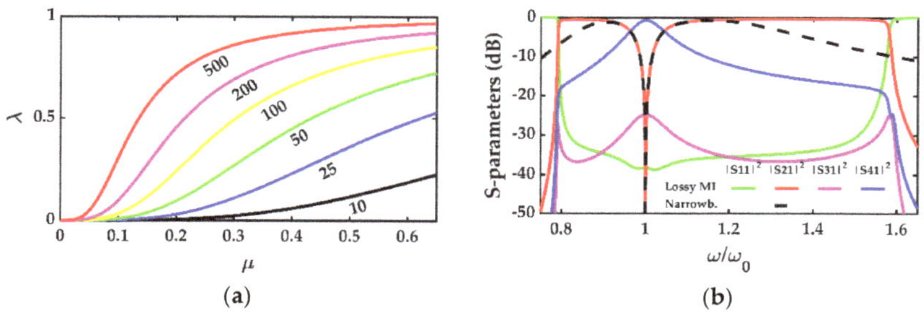

Figure 4. (**a**) Variation of λ with μ for different unloaded resonator Q-factors with $\kappa = 0.6$. (**b**) Frequency dependence of scattering parameters for infinite rejection filter with lossy MI and resistive narrowband terminations with $\kappa = 0.6$, $\kappa_c = 0.3$, and $Q = 200$.

2.4.2. Multiple Bandstop Frequencies

A directional filter with N complementary bandstop and bandpass responses can be realised by increasing the number of coupling resonators from 1 to N, as shown in Figure 5a. Here MI waveguides with coupling coefficients $\pm\kappa$ are connected using vertical lines containing N weakly coupled resonators with coupling κ_c. The resonator rows are labelled from 0 (bottom) to top $(N-1)$ and have the currents K_m and P_m flowing in the left and right columns, respectively. The lines of coupling resonators can be considered as two identical uniform MI waveguides with a passband of $\frac{1}{\sqrt{1+|\kappa_c|}} \leq \frac{\omega}{\omega_0} \leq \frac{1}{\sqrt{1-|\kappa_c|}}$ and dispersion relation $Z + Z_{mc}(e^{-jk_c a} + e^{jk_c a}) = 0$, where k_c is the propagation constant. The currents P_m and K_m may also be expressed in terms of reflected and transmitted waves as:

$$P_m = P_T e^{-jk_c ma} + P_R e^{jk_c ma}, \quad K_m = K_T e^{-jk_c ma} + K_R e^{jk_c ma}. \tag{22}$$

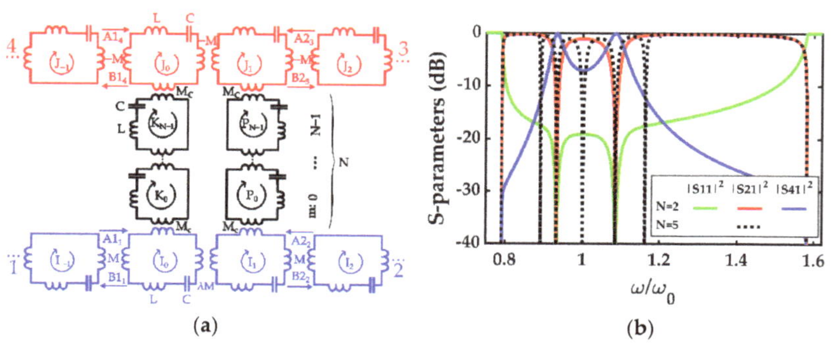

Figure 5. (**a**) Multiple bandstop MI directional filter; (**b**) frequency dependence of S-parameters for multiple passband filter with lossless MI terminations (1) for $\kappa = 0.6$, $\kappa_c = 0.3$, $N = 2$, and $N = 5$.

The solutions for currents I_n and J_n remain unchanged from Equation (8). Once again, we may write down the equations at the junctions as:

$$\begin{gathered} ZI_0 + Z_m(I_{-1} + I_1) + Z_{mc}K_0 = 0,\ ZI_1 + Z_m(I_0 + I_2) + Z_{mc}P_0 = 0 \\ ZK_0 + Z_{mc}(I_0 + K_1) = 0,\ ZP_0 + Z_{mc}(I_1 + P_1) = 0 \\ ZK_{N-1} + Z_{mc}(K_{N-2} + J_0) = 0,\ ZP_{N-1} + Z_{mc}(P_{N-2} + J_1) = 0 \\ ZJ_0 - Z_m(J_{-1} + J_1) + Z_{mc}K_{N-1} = 0,\ ZJ_1 - Z_m(J_0 + J_2) + Z_{mc}P_{N-1} = 0. \end{gathered} \tag{23}$$

Equation (23) can again be solved for the reflection coefficient at port 1 and transmission coefficients at ports 2–4 by substituting the dispersion relations and eliminating the terms P_T, P_R, and K_T, K_R. When $N = 1$, Equation (24) is identical to Equation (11).

$$\begin{gathered} \tfrac{R}{T} = (e^{j2k_c a(N+1)} - 1)\left\{e^{j2k_c a}\left(1 + e^{j2k_f a}\right) - \mu e^{j(k_f+k_c)a}\left(1 + e^{j2k_c a}\right) + \mu^2 e^{j2(k_f+k_c)a}\right\} / D \\ \tfrac{T_2}{T} = \left(1 - e^{j2(N+1)k_c a}\right)\left\{2e^{j2(k_f+k_c)a} - \mu e^{j(3k_f+k_c)a}\left(1 + e^{j2k_c a}\right)\right\} / D \\ \tfrac{T_4}{T} = \mu e^{j(N+1)k_c a}\left\{2e^{j(k_f+k_c)a}\left(1 - e^{j2k_c a}\right) - \mu e^{j2k_f a}\left(1 - e^{j4k_c a}\right)\right\} / D \\ D = \left(e^{j2ka} - 1\right)e^{j2k_c a}\left(1 - e^{j2k_c a(N+1)}\right) + 2\mu e^{j(k_f+k_c)a}\left(1 - e^{j2k_c a(N+2)}\right) - \mu^2 e^{j2ka}\left(1 - e^{j2k_c(N+3)a}\right) \end{gathered} \tag{24}$$

The coefficients can be converted to scattering parameters as before. Figure 5b shows the frequency dependence of S-parameters for $N = 2$ and $N = 5$. The additional coupling resonators generate multiple bandpass responses in S_{41} and complementary notches in S_{21}. The value of $|\kappa_c|$ determines the sharpness of the peaks in S_{41}, as well as the notch frequencies, with a larger separation between peaks for higher values. The quality factors of the peaks are higher for increasing N, and the effect of loss is to reduce rejection in S_{21} and increase insertion loss in S_{41}.

The frequencies of peaks and notches can be found by computing the maxima of S_{41} or the minima of S_{21} within the MI passband. The central frequencies ω_k simplify to:

$$\omega_k = \omega_0\sqrt{\frac{1}{1 + \kappa_c \cos\left(\frac{\nu\pi}{N+1}\right)}},\ \nu \in [1, N]. \tag{25}$$

The notch spacing is nonuniform and can be adjusted by varying ω_0 or κ_c. The values for ω_k are independent of κ, provided that $\kappa > \kappa_c$. When N is odd, there is always a peak at the resonant frequency. As previously shown, infinite rejection can be achieved at resonance for $N = 1$. However, since path losses cannot be equalised at several frequencies, infinite rejection cannot be obtained for all notches simultaneously when $N > 1$. However, simulations show that it is simple to improve performance by adjusting λ.

2.4.3. Three-Port Filters and Multiplexers

In the absence of infinite or improved rejection ($\lambda = 1$), the current in port 3 is identically zero at all frequencies with port 1 as an input. This implies that port 3 can be physically removed from either the single- or multiple-notch device, along with a resonator at n=1 in the backward-wave guide in Figure 2a or Figure 5a. Hence, a three-port directional filter can be generated with the addition of just three weakly coupled resonators, and a backward-wave resonator at $n = 0$ loaded with one of the terminations from Figure 2b. Loss compensation, however, is then impossible due to the removal of signal paths. Nevertheless, filter performance would be indistinguishable from Figure 3a or Figure 5b, as long as λ is unity. Multiplexers can also be implemented by cascading filters, with the notch frequencies of the constituent filters defining the frequencies of extracted signals.

3. Results and Discussion

In this section, we present experimental results for filters constructed for $f_0 = 13.56$ MHz to remove residual carrier from tag responses in HF RFID. A tuneable design was developed to allow the demonstration of filter operation and extension to complex filters.

3.1. Single-Notch Filter

The overall design was as shown in Figure 2a; however, resistive terminations were used, and ports were floated using paired capacitors. To allow frequency tuneability, all capacitors included varicaps. To confine magnetic fields and ensure nearest-neighbour coupling, all inductors used low-loss ferrimagnetic toroidal cores with inner and outer diameters of 13.2 and 21 mm (Fair-Rite 5967000601, based on "67" Ni–Zn ferrite material with cutoff at 50 MHz). The initial relative permeability was ≈40, implying that the magnetic flux was almost completely contained in the core. Two windings (with inductances L_C and L_T, such that $L_C + L_T = L$) were used for the waveguide elements, and three (with inductances L_C, L_C, and L_S, such that $2L_C + L_S = L$) for the coupling resonators. Adjacent windings were reversed to obtain negative coupling coefficients. A method was used to allow the tuning of mutual inductance using components constructed by 3D printing with a polycarbonate filament. Figure 6a shows this mechanism, allowing control over the relative position of two copper windings. The coupling coefficient was a nonlinear function of the winding separation with minimum separation yielding maximum coupling.

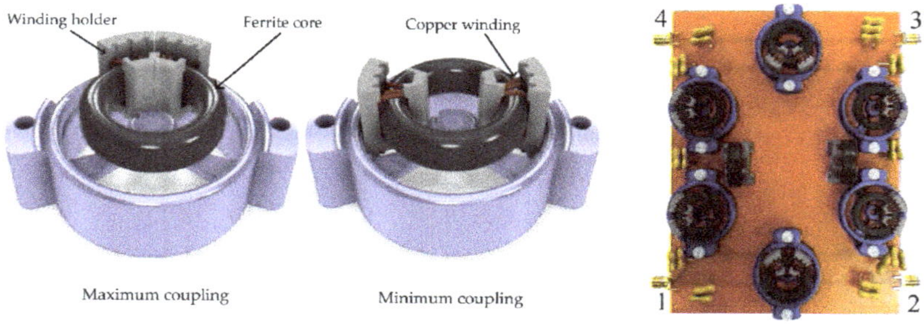

(a) (b)

Figure 6. MI directional filter: (**a**) mechanism for tuning of magnetic coupling (LH–maximum coupling; RH–minimum coupling); (**b**) complete PCB with labelled port numbers.

Figure 6b shows a photograph of the completed filter constructed as a copper-clad printed circuit board using low-loss mica capacitors. Mechanical tuning was provided for all mutual inductances. The value of M needed for impedance matching to a 50 Ω load at resonance was first identified. Inductors L_s were then chosen to equalise the self-inductance of all resonators, and capacitances were tuned to achieve resonance at 13.56 MHz. Table 1 shows the values of the most important parameters after tuning.

Table 1. Experimental parameter values.

Parameter	Value	Range of Adjustment
L_T	1.16 µH	N/A (not applicable)
L_c	460 nH	N/A
L_s	710 nH	N/A
κ	0.725	0.51–0.92
κ_c	0.28	0.19–0.34
ω_0	13.56 MHz	DC to 50 MHz
Q	200 ± 10%	N/A

Four-port scattering parameters were measured using an electronic network analyser (Agilent E5061B). To begin with, λ was set to unity so that responses were independent of the port used as input. Figure 7a compares the frequency dependence of S-parameters with the predictions of the theoretical model for $\kappa_c = 0.28$. Excellent agreement with experimental results can be seen. Larger values of μ were investigated and found to reduce losses and increase the passband of the bandpass channel and the notch depth.

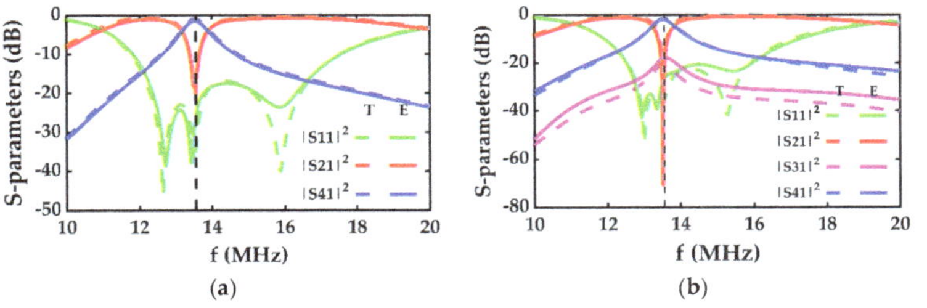

Figure 7. Experimental (solid) and theoretical (dashed) frequency dependence of scattering parameters for MI filter with resistive narrowband terminations (2) with $\omega_0 = 13.56$ MHz, $Q = 200$, and $Z_{0M} = 50\ \Omega$. (**a**) $\kappa_c = 0.28$ and $\lambda = 1$; (**b**) infinite rejection with $\kappa_c = 0.25$ and $\lambda = 0.79$.

3.2. Single-Notch Filter with Infinite Rejection

Stopband rejection can be greatly improved by fine-tuning λ according to Equation (21). For parameters in Table 1, the value of λ required for infinite rejection is 0.79, well within range for the coupling tuning mechanism. Figure 7b compares the simulated and measured scattering parameters obtained in this case. A performance improvement of over 50 dB is achieved compared with that in Figure 7a, with 70 dB rejection at port 2. This improvement is at the expense of reduced isolation of port 3 with a minimum attenuation of 18 dB at resonance between ports 1 and 3. However, losses in S_{21} are still extremely low.

3.3. Double-Notch Filters

A two-notch directional filter with parameter values shown in Table 1 was implemented in the same way by increasing the number of coupling resonators in each vertical branch from one to two. The magnitude of κ_c determines the frequency locations of the notches in S_{21} according to Equation (25). However, smaller values of κ_c also increase propagation losses through the coupling resonators, reducing notch depth and increasing losses in the bandpass channel. Figure 8a compares theoretical and experimental results for $\kappa_c = 0.3$. Agreement between simulation and experimental results is again excellent. Attenuation in S_{21} is the same as for the single-notch filter, implying that two signals can be extracted simultaneously without hindering performance.

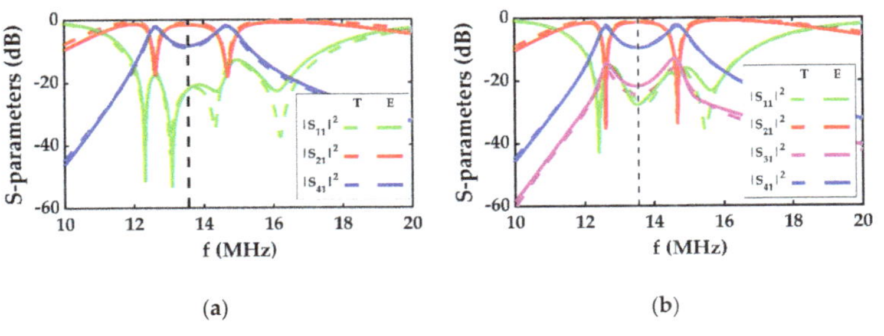

Figure 8. Experimental (solid) and theoretical (dashed) frequency dependence of S-parameters for double passband MI filter with resistive narrowband terminations with $\omega_0 = 13.56$ MHz, $Q = 180$, $Z_{0M} = 50\ \Omega$, and $\kappa_c = 0.3$. (**a**) No improved rejection; (**b**) improved rejection.

3.4. Double-Notch Filter with Improved Rejection

Although infinite rejection appears impossible in multinotch filters, rejection can be improved by adjusting λ. Figure 8b compares theoretical and measured S-parameters for a two-notch filter with $\kappa_c = 0.3$ with λ chosen to equalise rejection at the two notches. A significant performance improvement in a rejection of over 15 dB is obtained, once again at the expense of port 3 isolation. As regards single-notch filters, lower values of λ are needed for higher values of μ. Low insertion loss is observed around the resonance frequency but increases elsewhere as matching between the guides and load deteriorates.

3.5. Applications

Toroid transformers based on soft ferrites allow magnetoinductive designs to be realised with low non-nearest-neighbour coupling. However, core materials must have low loss and high linearity. Some applications also require high power handling. In HF RFID, a modulated AC magnetic field delivers power and data to tags, which couple to the field using inductive loops. For the ISO/IEC 14443 type A standard, the carrier frequency is $f_c = 13.56$ MHz, while tags use an internally generated subcarrier at $f_m = \frac{f_c}{16}$ to return data, generating sideband spectra with central frequencies, $f_c \pm f_m$. Since the carrier power can be orders of magnitude greater than that of the sidebands, receiver protection is required. The single-notch filter presented here can be integrated into existing MI systems via direct connection to the waveguides, has high rejection at resonance, and may be used to transfer residual carrier to a load. However, the filter must be effective at the representative carrier powers, despite potential nonlinearities introduced by ferrite cores. To emulate generic HF RFID signals, we modulated a high-power carrier at f_c using a simple cosine message waveform at f_m. The waveform of the input signal $s(t)$ is:

$$s(t) = [A_c + A_m \cos(2\pi f_m t)] \cos(2\pi f_c t). \tag{26}$$

Here A_c and A_m are the carrier and message amplitudes. The signal was derived from a first signal generator (Agilent E4433B) modulated by a second signal modulator (Agilent N5181A) with amplification by a 27 dB power amplifier (Minicircuits ZFDWHA-1–20+). A_c was adjusted to deliver a carrier power, P_c, into a 50 Ω load, while the modulation index A_m/A_c was chosen to generate typically small sidebands. Detection was performed using a digital storage oscilloscope (Keysight InfiniiVision DSOX3024T, Santa Rosa, CA, USA) using a 30 dB attenuator for unfiltered signals. Figure 9a shows the power spectral density (PSD) of the signal input to port 1 of the single-notch filter for $P_c = 1$ W. Amplifier nonlinearities have resulted in minor harmonics of the AM spectrum, which have been partially suppressed at high frequencies with a low-pass filter. Figure 9b shows the PSD at port 2, the stopband channel. The output shows a carrier attenuation of 73 dB and a sideband attenuation of 1.2 dB, implying an output signal with almost intact message and negligible carrier. No significant additional harmonics are introduced. Figure 9c shows the attenuation of the two sidebands and the carrier over a power range between 0.1 and 1 W. Attenuation is almost independent of P_c, implying effective power handling despite the use of ferrites.

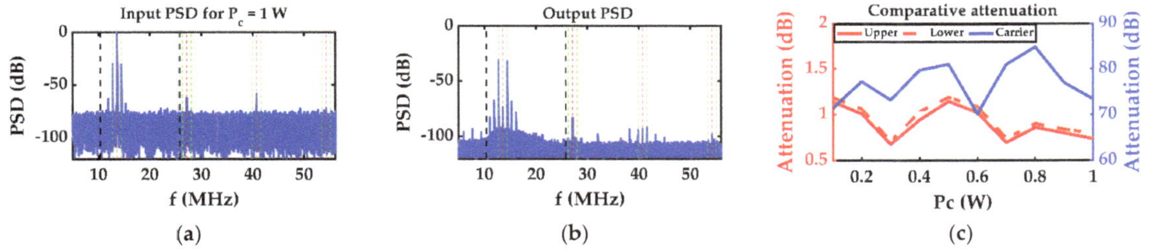

Figure 9. Power spectral densities of (**a**) input signal to port 1 of MI single-notch directional filter; (**b**) output signal at port 2 normalised by the input carrier power P_c. Dashed lines: filter bandwidth; red: harmonics at multiples of carrier frequencies; green: harmonics at message sidebands. (**c**) Attenuation of sidebands and carrier.

3.6. Frequency Scaling

The potential for scaling to higher frequencies is largely determined by ferrite performance. Table 2 shows performance parameters for low-cost NiZn soft ceramic ferrite materials and toroids designed for RF applications, such as EMI suppressors (upper group) and broadband transformers and baluns (lower), including their initial relative permeability μ_i, Curie temperature T_C, and loss factor. The main difference between groups is the maximum usable frequency F_{max}. Following Snoek's limit [52], μ_i falls as F_{max} is raised. High-frequency performance is therefore accompanied by a reduction in magnetic field confinement and an increase in loss. Small toroids may be used to minimise the volume of magnetic material. However, careful design will be required to optimise performance, power handling will be reduced, and automated coil winding may be required for construction [53]. Despite this, the data suggest that response may be extended to the low UHF band. Above this, air-cored inductors may be used, but non-nearest-neighbour coupling and radiation will inevitably complicate the design and reduce performance.

Table 2. Properties of a selection of ferrite materials and toroids.

Manufacturer	Model	OD (mm)	ID (mm)	Height (mm)	Initial μ_i	F_{max} (MHz)	T_C °C	Material	Loss Factor
Ferroxcube	-	-	-	-	125	20	>350	4C65	130 @ 10 MHz
Fair-Rite	5961001801	22.10	13.70	06.35	125	25	>300	61	10 @ 10 MHz
Ferroxcube	-	-	-	-	25	100	>400	4E2	-
Fair-Rite	5968001801	22.10	13.70	6.35	16	150	>500	68	300 @ 100 MHz
National Magnetics	-	-	-	-	7.5	400	>320	M5	<3500 @ 100 MHz

3.7. Future Work

MI directional filters can also be designed for use at UHF. Potential research could concentrate on counteracting ferrite losses or, alternatively, minimising non-nearest-neighbour coupling and radiation in open-loop designs. Furthermore, the device wave operating principles can be extended to other metamaterial types, such as elastic or acoustic [54]. In particular, it has already been shown by us that MEMS could be used to generate high-performance two-port notch and comb filters [55]. Once again, wave analysis was used to simplify the coupled dynamical equations. Extension to multiport devices can be a fitting research target. Higher performance is expected due to inherently higher Q-factors of MEMS resonators.

4. Conclusions

New configurations of a magnetoinductive device with directional filter properties have been introduced based purely on magnetically coupled LC resonators. Design rules have been established, and methods for calculating scattering parameters when filters are used in MI systems (which have complex, frequency-dependent impedance) or conventional systems (with real impedance) have been clarified. The circuit equations have been simplified and solved using wave analysis. Analytic expressions have been developed for S-parameters and bandwidth, a simple method for introducing tuneability has been proposed, and extensions to allow infinite rejection at a chosen frequency or multiple stopbands with high rejection have been described.

Device construction and operation have been verified at HF. Ceramic ferrite cores have been used to ensure nearest-neighbour coupling in a compact layout, and simple mechanical methods of tuning resonant frequency and coupling (essential in blocker applications) have been proposed and demonstrated. Experimental results for tuneable filters, filters with infinite rejection, and filters with multiple bandstop frequencies have all been shown to agree with theoretical models. Power handing capability sufficient for blocker applications in HF RFID has been demonstrated, with little harmonic generation due to ferrite nonlinearity. Further development will involve direct connection to a magnetoinductive antenna, and this work is in progress.

Author Contributions: Methodology, A.V., R.R.A.S. and O.S.; software, A.V.; validation, A.V., R.R.A.S. and O.S.; formal analysis, A.V., R.R.A.S. and O.S.; investigation, A.V.; writing—original draft preparation, A.V., R.R.A.S. and O.S.; writing—review and editing, A.V., R.R.A.S. and O.S. All authors have read and agreed to the published version of the manuscript.

Funding: This research received no external funding.

Conflicts of Interest: The authors declare no conflict of interest.

References

1. Cohn, S.; Coale, F. Directional Channel-Separation Filters. *Proc. IRE* **1956**, *44*, 1018–1024. [CrossRef]
2. Coale, F.S. A Traveling-Wave Directional Filter. *IRE Trans. Micr. Theory Tech.* **1956**, *4*, 256–260. [CrossRef]
3. Walker, J.L.B. Exact and Approximate Synthesis of TEM-Mode Transmission-Type Directional Filters. *IEEE Trans. Micr. Theory Tech.* **1978**, *26*, 186–192. [CrossRef]
4. Matthaei, G.L.; Young, L.; Jones, E.M. *Design of Microwave Filters, Impedance-Matching Networks, and Coupling Structures*; McGraw-Hill: New York, NY, USA, 1963; Chapter 14.
5. Coale, F.S. Applications of Directional Filters for Multiplexing Systems. *IEEE Trans. Micr. Theory Tech.* **1958**, *6*, 450–453. [CrossRef]
6. Wing, O. Cascade Directional Filter. *IEEE Trans. Micr. Theory Tech.* **1959**, *7*, 197–201. [CrossRef]
7. Cameron, R.; Yu, M. Design of manifold-coupled multiplexers. *IEEE Microw. Mag.* **2007**, *8*, 46–59. [CrossRef]
8. Tuttle, L.P.; Wanselow, R.D. Practical Design of Strip-Transmission-Line Half-Wavelength Resonator Directional Filters. *IEEE Trans. Microw. Theory Tech.* **1959**, *7*, 168–173. [CrossRef]
9. Moore, B.D.; Cogdell, J.R. A Millimeter-Wave Directional Filter Cavity. *IEEE Trans. Microw. Theory Tech.* **1976**, *24*, 843–847. [CrossRef]
10. Uysal, S. Microstrip loop directional filter. *Electron. Lett.* **1997**, *33*, 475. [CrossRef]
11. Rosloniec, S.; Habib, T. Novel microstrip-line directional filters. *IEEE Trans. Microw. Theory Tech.* **1997**, *45*, 1633–1637. [CrossRef]
12. Cheng, Y.; Hong, W.; Wu, K. Half Mode Substrate Integrated Waveguide (HMSIW) Directional Filter. *IEEE Microw. Wireless Comp. Lett.* **2007**, *17*, 504–506. [CrossRef]
13. Kim, J.P. Improved Design of Single-Section and Cascaded Planar Directional Filters. *IEEE Trans. Microw. Theory Tech.* **2011**, *59*, 2206–2213. [CrossRef]
14. Zhang, Y.; Shi, S.; Martin, R.D.; Prather, D.W. Slot-Coupled Directional Filters in Multilayer LCP Substrates at 95 GHz. *IEEE Trans. Microw. Theory Tech.* **2017**, *65*, 476–483. [CrossRef]
15. Zhang, Y.-B.; Chen, B.; Ran, C.-Z. An X-band single-layer waveguide directional filter with compact size and low insertion loss. *IEICE Electron. Express* **2018**, *15*, 20180826. [CrossRef]
16. Stone, A.M.; Lawson, J.L. Infinite-Rejection Filters. *J. Appl. Phys.* **1947**, *18*, 691–703. [CrossRef]
17. Parthasarathy, D.; Harjani, R. Novel integratable notch filter implementation for 100 dB image rejection. In Proceedings of the 2003 International Symposium on Circuits and Systems, ISCAS'03, Bangkok, Thailand, 25–28 May 2003.
18. Jachowski, D.R. Passive enhancement of resonator Q in microwave notch filters. In Proceedings of the 2004 IEEE MTT-S International Microwave Symposium Digest (IEEE Cat. No.04CH37535), Fort Worth, TX, USA, 6–11 June 2004.
19. Jachowski, D.R. Compact, frequency-agile, absorptive bandstop filters. In Proceedings of the IEEE MTT-S International Microwave Symposium Digest, Long Beach, CA, USA, 17 June 2005.
20. Ghaffari, A.; Klumperink, E.A.M.; Nauta, B. Tunable N-Path Notch Filters for Blocker Suppression: Modeling and Verification. *IEEE J. Solid-State Circuits* **2013**, *48*, 1370–1382. [CrossRef]

21. Lababidi, R.; Le Roy, M.; Le Jeune, D.; Perennec, A.; Vauche, R.; Bourdel, S.; Gaubert, J. Compact highly selective passive notch filter for 3.1–5 GHz UWB receiver system. In Proceedings of the 2015 IEEE International Conference on Electronics, Circuits, and Systems (ICECS), Cairo, Egypt, 6 December 2015.
22. Morgan, M.A.; Boyd, T.A. Theoretical and Experimental Study of a New Class of Reflectionless Filter. *IEEE Trans. Microw. Theory Tech.* **2011**, *59*, 1214–1221. [CrossRef]
23. Morgan, M.A.; Boyd, T.A. Reflectionless Filter Structures. *IEEE Trans. Microw. Theory Tech.* **2015**, *63*, 1263–1271. [CrossRef]
24. Standley, R.D. A Time-Delay Equalizer Using Directional Filter Cascades. *IEEE Trans. Microw Theory Tech.* **1971**, *19*, 497–498. [CrossRef]
25. Erickson, N.R. A Directional Filter Diplexer Using Optical Techniques for Millimeter to Submillimeter Wavelengths. *IEEE Trans. Microw. Theory Tech.* **1977**, *25*, 865–866. [CrossRef]
26. Hunter, I.; Musonda, E.; Parry, R.; Guess, M.; Meng, M. Transversal directional filters for channel combining. In Proceedings of the 2013 IEEE MTT-S International Microwave Symposium Digest (MTT), Seattle, WA, USA, 2–7 June 2013.
27. Sorocki, J.; Piekarz, I.; Gruszczynski, S.; Wincza, K. Miniaturized directional filter multiplexer for band separation in UWB antenna systems. In Proceedings of the 2015 International Symposium on Antennas and Propagation (ISAP), Hobart, Australia, 9–12 November 2015; pp. 1–4.
28. Wincza, K.; Gruszczynski, S. Frequency-Selective Feeding Network Based on Directional Filter for Constant-Beamwidth Scalable Antenna Arrays. *IEEE Trans. Antennas Propag.* **2017**, *65*, 4346–4350. [CrossRef]
29. Sun, J.S.; Lobato-Morales, H.; Choi, J.H.; Corona-Chavez, A.; Itoh, T. Multistage Directional Filter Based on Band-Reject Filter with Isolation Improvement Using Composite Right-/Left-Handed Transmission Lines. *IEEE Trans. Microw. Theory Tech.* **2012**, *60*, 3950–3958. [CrossRef]
30. Beruete, M.; Falcone, F.; Freire, M.J.; Marqués, R.; Baena, J.D. Electroinductive waves in chains of complementary metamaterial elements. *Appl. Phys. Lett.* **2006**, *88*, 083503. [CrossRef]
31. Liu, N.; Kaiser, S.; Giessen, H. Magnetoinductive and Electroinductive Coupling in Plasmonic Metamaterial Molecules. *Adv. Mater.* **2008**, *20*, 4521–4525. [CrossRef]
32. Navarro-Cia, M.; Carrasco, J.M.; Beruete, M.; Falcone, F.J. Ultra-wideband metamaterial filter based on electro-inductive wave coupling between microstrips. *PIER Lett.* **2009**, *12*, 141–150. [CrossRef]
33. Gil, M.; Velez, P.; Aznar-Ballesta, F.; Munoz-Enano, J.; Martin, F. Differential Sensor Based on Electroinductive Wave Transmission Lines for Dielectric Constant Measurements and Defect Detection. *IEEE Trans. Antennas Propag.* **2020**, *68*, 1876–1886. [CrossRef]
34. Shamonina, E.; Kalinin, V.A.; Ringhofer, K.H.; Solymar, L. Magneto-inductive waveguide. *Electron. Lett.* **2002**, *38*, 371. [CrossRef]
35. Shadrivov, I.V.; Reznik, A.N.; Kivshar, Y.S. Magnetoinductive waves in arrays of split-ring resonators. *Physica B* **2007**, *394*, 180–183. [CrossRef]
36. Stevens, C.J.; Chan, C.W.T.; Stamatis, K.; Edwards, D.J. Magnetic Metamaterials as 1-D Data Transfer Channels: An Application for Magneto-Inductive Waves. *IEEE Trans. Microw. Theory Tech.* **2010**, *58*, 1248–1256. [CrossRef]
37. Sun, Z.; Akyildiz, I.F. Magnetic Induction Communications for Wireless Underground Sensor Networks. *IEEE Trans. Antennas Propag.* **2010**, *58*, 2426–2435. [CrossRef]
38. Gulbahar, B.; Akan, O.B. A Communication Theoretical Modeling and Analysis of Underwater Magneto-Inductive Wireless Channels. *IEEE Trans. Wireless Comms* **2012**, *11*, 3326–3334. [CrossRef]
39. Zhong, W.; Lee, C.K.; Hui, S.Y.R. General Analysis on the Use of Tesla's Resonators in Domino Forms for Wireless Power Transfer. *IEEE Trans. Indust. Electron.* **2013**, *60*, 261–270. [CrossRef]
40. Agbinya, J.I. A magneto-inductive link budget for wireless power transfer and inductive communication systems. *PIER C* **2013**, *37*, 15–28. [CrossRef]
41. Stevens, C.J. A magneto-inductive wave wireless power transfer device. *Wireless Power Transf.* **2015**, *2*, 51–59. [CrossRef]
42. Freire, M.J.; Marqués, R. Planar magnetoinductive lens for three-dimensional subwavelength imaging. *Appl. Phys. Lett.* **2005**, *86*, 182505. [CrossRef]
43. Syms, R.R.A.; Young, I.R.; Ahmad, M.M.; Rea, M. Magnetic resonance imaging using linear magneto-inductive waveguides. *J. Appl. Phys.* **2012**, *112*, 114911. [CrossRef]
44. Floume, T. Magneto-inductive conductivity sensor. *Metamaterials* **2011**, *5*, 206–217. [CrossRef]
45. Yan, J.; Stevens, C.J.; Shamonina, E. A Metamaterial Position Sensor Based on Magnetoinductive Waves. *IEEE Open J. Antennas Propag.* **2021**, *2*, 259–268. [CrossRef]
46. Syms, R.R.A.; Sydoruk, O.; Wiltshire, M.C.K. Magneto-Inductive HF RFID System. *IEEE J. RFID* **2021**, *5*, 148–153. [CrossRef]
47. Shamonina, E.; Solymar, L. Magneto-inductive waves supported by metamaterial elements: Components for a one-dimensional waveguide. *J. Phys. D Appl. Phys.* **2004**, *37*, 362–367. [CrossRef]
48. Syms, R.R.A.; Shamonina, E.; Solymar, L. Magneto-inductive waveguide devices. *IEE Proc. Microw. Antennas Propag.* **2006**, *153*, 111. [CrossRef]
49. Syms, R.R.A.; Solymar, L.; Young, I.R. Broadband coupling transducers for magneto-inductive cables. *J. Phys. D Appl. Phys.* **2010**, *43*, 285003. [CrossRef]
50. Kurokawa, K. Power Waves and the Scattering Matrix. *IEEE Trans. Microw. Theory Tech.* **1965**, *13*, 194–202. [CrossRef]
51. Voronov, A.; Sydoruk, O.; Syms, R.R.A. Power waves and scattering parameters in magneto-inductive systems. *AIP Adv.* **2021**, *11*, 045327. [CrossRef]

52. Snoek, J.L. Dispersion and absorption in magnetic ferrites at frequencies above one Mc/s. *Physica* **1948**, *14*, 207–217. [CrossRef]
53. Kazimierczuk, M. *High-Frequency Magnetic Components*; John Wiley & Sons: Chichester, UK, 2014. [CrossRef]
54. Wang, Y.; Wang, Y.; Wu, B.; Chen, W.; Wang, Y. (February 26, 2020). Tunable and active phononic crystals and metamaterials. *ASME. Appl. Mech. Rev.* **2020**, *72*, 040801. [CrossRef]
55. Bouchaala, A.; Syms, R.R.A. New architectures for micromechanical coupled beam array filters. *Microsyst. Technol.* **2021**, *27*, 3377–3387. [CrossRef]

electronics

MDPI

Article

A New Reconfigurable Filter Based on a Single Electromagnetic Bandgap Honey Comb Geometry Cell

Andre Tavora de Albuquerque Silva [1,2,*], Claudio Ferreira Dias [1,*], Eduardo Rodrigues de Lima [1], Gustavo Fraidenraich [2,*] and Larissa Medeiros de Almeida [3]

[1] Instituto de Pesquisas Eldorado, Campinas 13083-898, Brazil; eduardo.lima@eldorado.org.br
[2] DECOM/FEEC/UNICAMP, State University of Campinas, Campinas 13083-852, Brazil
[3] Transmissora Aliança de Energia Elétrica S.A., Rio de Janeiro 20010-010, Brazil; larissa.almeida@taesa.com.br
* Correspondence: andre.albuquerque@eldorado.org.br (A.T.d.A.S.); claudio.dias@eldorado.org.br (C.F.D.); gf@decom.fee.unicamp.br (G.F.)

Abstract: This work presents a new unit cell electromagnetic bandgap (EBG) design based on HoneyComb geometry (HCPBG). The new HCPBG takes a uniplanar geometry (UCPBG—uniplanar compact PBG) as a reference and follows similar design methods for defining geometric parameters. The new structure's advantages consist of reduced occupied printed circuit board area and flexible rejection band properties. In addition, rotation and slight geometry modification in the HCPBG cell allow changing the profile of the attenuation frequency range. This paper also presents a reconfigurable unit cell HCPBG filter strategy, for which the resonance center frequency is shifted by changing the gap capacitance with the assistance of varactor diodes. The HCPBG filter and reconfiguration behavior is demonstrated through electromagnetic (EM) simulations over the FR1 band of the 5G communication network. Intelligent communication systems can use the reconfiguration feature to select the optimal operating frequency for maximum attenuation of unwanted or interfering signals, such as harmonics or intermodulation products.

Keywords: electromagnetic bandgap; photonic bandgap; electromagnetic compatibility; interference; filtering; HCPBG (HoneyComb PBG)

Citation: Tavora de Albuquerque Silva, A.; Ferreira Dias, C.; Rodrigues de Lima, E.; Fraidenraich, G.; Medeiros de Almeida, L. A New Reconfigurable Filter Based on a Single Electromagnetic Bandgap Honey Comb Geometry Cell. *Electronics* **2021**, *10*, 2390. https://doi.org/10.3390/electronics10192390

Academic Editors: Giovanni Andrea Casula and Naser Ojaroudi Parchin

Received: 12 July 2021
Accepted: 20 September 2021
Published: 30 September 2021

1. Introduction

Photonic bandgaps (PBGs) are periodic structures that introduce material changes in the waveguide or printed circuit board (PCB), such as holes, patterns, or dielectric rods. They are also known as photonic crystals and are based on Electromagnetic Band Gap (EBG) properties. The geometric modification in the medium produces forbidden frequency bands for propagating waves. In other words, it works as a wave filter with high impedance at desired frequency bands. As a result, the electromagnetic waves in a PBG material are hindered due to the periodic discontinuity, which is equivalent to a photonic crystal in the light domain. They were first reported for structures at optical wavelengths in 1987 [1,2]. Since the first PBG structure publication, designers created many shapes, geometries, and materials for various applications. As an example, we can cite: microwave filters, electromagnetic compatibility (EMC) improvement [3–5], antenna beam steering [6], compact antenna arrays [7], antenna gain, efficiency, and bandwidth improvement [8–10], and beam tilting of 5G antenna arrays [11].

PBG structures have the characteristic of phase control of plane waves enabling suppression of surface waves and higher-order harmonics. It is also an exciting tool for those looking beyond 5G, where the advent of electromagnetic components can shape how they interact with the propagation environment. It is a case for reconfigurable intelligent surface (RIS), a two-dimensional surface of engineered material whose properties are reconfigurable rather than static [8,9,12]. For example, the RIS device can control scattering, absorption, reflection, and diffraction properties by software according to how the environment changes over time. In principle, the RIS can form a beam and synthesize the scattering behavior of an arbitrarily shaped surface of the same size. For example, it can create a superposition of multiple beams or act as a diffuse scatterer [13]. Regarding traditional communication systems, the critical difference between a RIS and the common notion of an antenna array in 5G is that a RIS is neither part of the transmitter nor the receiver. However, it is a controllable part of the wireless propagation environment. The system-level role of a RIS is to influence the propagation of the wireless signals sent by other devices without generating its signals [14].

On the scope of the current investigation, we focus on interference filtering, which is a critical problem to be addressed as the number of wireless devices increases every year. It is of great concern that long-term interfering issues could arise due to the increasing number of heterogeneous devices. Aiming on that, it is a necessity to implement such solutions as a filtering option. For example, cable-stayed power line towers monitored by wireless sensors operating at unlicensed spectrum deliver data of utmost importance to the substations. The data help predict the risks of collapse events, and wireless sensor communication may be harmed by adjacent heterogeneous systems interference. In this way, one can use PBG as a tool to contend against interference coming from a crowded environment in several ways. Firstly, the current downscaling of devices makes structure sizes a critical issue. As technologies occupy higher spectrum frequencies, PBG is a viable tool to be adopted since the stopband frequency is proportional to half of the guided wavelength [15]. Secondly, planar PBGs like UCPBG do not need vias or unique materials suitable for standard PCB designs with no extra cost. The UCPBG works as a stopband filter due to the structure of metal etched in slots on the ground plane connected by narrow lines to form a distributed LC network [16,17].

In this paper, we present the HCPBG unit cell, a novel geometry for EBG structures that introduces new features when compared to that of the traditional UCPBG [9]. For instance, it further decreases the consumption area compared to that of the UCPBG equivalent due to the proposed hexagonal geometry. Secondly, the hexagon can keep the same perimeter as the side of a square, where the microstrip line crosses the PBG structure while occupying 65% of the area. Besides, the HCPBG unit cell can be configured with different traces and orientations, adding asymmetry to the design and producing different filtering characteristics, as simulations and measurements demonstrate. As a result of this study, we can highlight the following HCPBG advantages:

- Introduction of a new EBG geometrical structure for RF filtering applications;
- A more effective area usage for PCBs due to the hexagon geometry;
- Multiple rejection frequency bands and bandwidths are achieved depending on the orientation or number of traces;
- Addition or suppression of traces allow shifting the first resonance frequency;
- Suppression of high-frequency rejection band by changing the orientation;
- Wide bandwidths are obtained by changing the number of traces or orientation;
- Active reconfiguration of the first resonance frequency.

As a second novel feature, we also present a reconfiguration method to modify the center frequency of the rejection band by controlling the gap capacitance through a varactor diode. The reverse voltage of the varactor diode changes its capacitance, allowing a control system to shift filtering center frequency and bandwidth range.

Although the technique of using active discrete components to reconfigure different types of PBG structures was studied before, the reconfiguration of the UCPBG type structures, as presented in this work, is not yet reported in the literature.

Different applications of reconfigurable EBG structures were investigated over the years, such as antenna array beam-steering at 6 GHz using metal tape [6], change of patch antenna polarization for navigation systems with the use of varactor diodes [18], modification of antenna frequency and radiation pattern for WiFi/WiMAX using PIN diode [19], cavity resonator at 10 GHz range with mechanical switching reconfiguration [20], and reconfigurable 3D MEMS filter for optical applications [21].

The HCPBG is a planar structure, and no unique material, vias, or fabrication process is needed to produce such structures. This is an advantage over tunable bandpass filters based on microelectromechanical systems (MEMS) [22]. On the other hand, planar reconfigurable PBG filters, as described in [23–25], have the advantage of large bandwidth and dynamic range. However, to achieve a frequency range of a few GHz, they show an increased area consumption and circuit complexity compared to that of the proposed filter.

This paper organizes as follows: Section 2 refers to the presented HCPBG model. Section 3 presents the simulation results comparing the classic UCPBG to the HCPBG and multiple geometries through transmission loss of a microstrip line over one single cell; it also presents the effect of different orientations. Section 4 shows the measurement setup and transmission loss results for the fabricated samples. Section 5 describes the design of the reconfigurable HCPBG filter and presents the simulation and measurement results. Finally, the discussion is closed in the Section 6.

2. Design of HCPBG

The method used to evaluate the UCPBG and HCPBG unit cells consists of a PCB having a microstrip line that crosses the entire top layer and one single cell at the bottom layer underneath the microstrip line. Figure 1 illustrates a diagram used in this work for simulation and measurements. By supposing an interfering signal impinging the stripline, the objective here is to calculate a resonance frequency for the structure to filter out the interference.

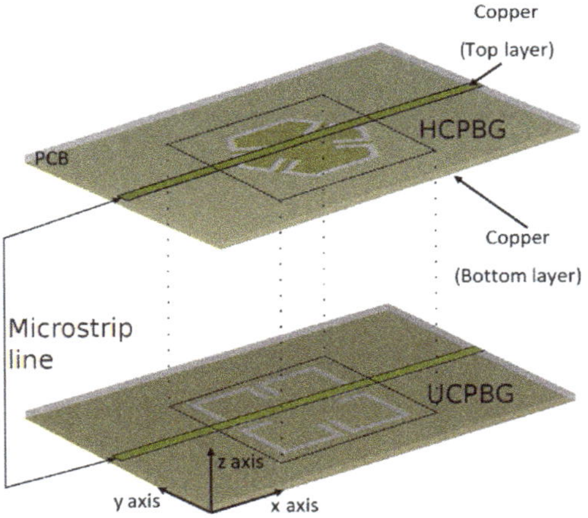

Figure 1. Reference diagram for simulations and measurements.

Figure 2 shows (a) the design reference UCPBG and (b) the new proposed HCPBG unit cells on microstrip substrates. The dark area is the metallic ground layer, while the white area represents the removal of the metal (air gaps). The geometry design parameters for both UCPBG and HCPBG are the gap size g_1, g_2, the total structure size a, and the trace dimensions where t_l is the length and t_w is the width. The gap sizes define the capacitance, while the trace length and width define the inductance. For example, reducing the gap size g_1, g_2 increases the capacitance and, henceforth, resonance moves to lower frequencies. On the other hand, a longer length t_l will produce a higher inductance, consequently reducing the rejection band frequency. Similarly, the width t_w affects the inductance and, henceforth, the resonance frequency.

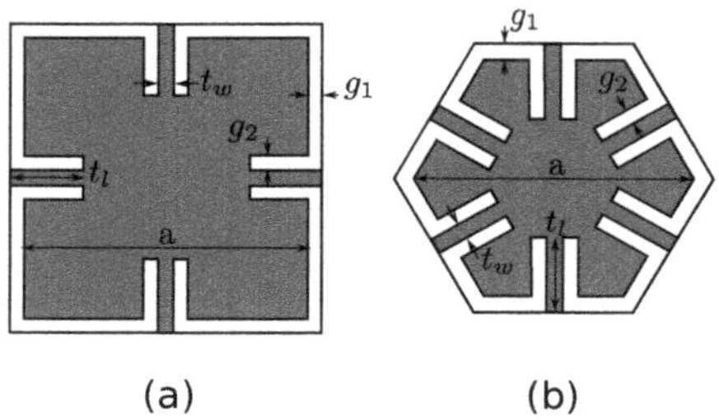

Figure 2. PBG structure designs—(**a**) Uniplanar Geometry PBG (UCPBG) and (**b**) HoneyComb PBG (HCPBG).

The lumped capacitors and inductors introduced by gaps and traces form a parallel LC network as described in [2]. The determination of a stopband frequency for a single PBG structure is roughly determined by [5]

$$f_o = \frac{1}{2\pi\sqrt{LC}}, \tag{1}$$

where L is the inductance of the trace and C is the capacitance of the gaps connecting the PBG structure to the ground. The method for calculating C follows the model of two metal sheets coplanar capacitance on a PCB [26], which can be calculated as

$$C = \begin{cases} \frac{l\varepsilon_r}{377\pi c}\ln\left[-\frac{2\left(1+\sqrt[4]{1-\frac{g_1^2}{(2w+g_1)^2}}\right)}{-1+\sqrt[4]{1-\frac{g_1^2}{(2w+g_1)^2}}}\right], 0 < \frac{g_1}{g_1+2w} \leqslant \frac{1}{\sqrt{2}}, \\ \frac{l\varepsilon_r}{120c\ln\left[-\frac{2\left(1+\sqrt{\frac{g_1}{2w+g_1}}\right)}{-1+\sqrt{\frac{g_1}{2w+g_1}}}\right]}, \frac{1}{\sqrt{2}} < \frac{g_1}{g_1+2w} \leqslant 1, \end{cases} \tag{2}$$

where w is the size of the metal plates of the considered capacitor model, $\varepsilon_r = 4.3$ (FR4) is the substrate dielectric constant, c = 299,792,458 m/s is the vacuum speed of light, and $l = a - t_w - g_1 - g_2$ is the gap length. It is also convenient to make $g_2 = t_w$ to minimize the effect on the inductance L.

One can calculate L as [27]

$$L = \rho(t_w) \times t_l, \tag{3}$$

where

$$\rho(t_w) = \begin{cases} 60\frac{\ln\left[\frac{8h}{t_w} + \frac{t_w}{4h}\right]}{c}, & \text{if } \frac{t_w}{h} \leq 1 \\ \frac{180h\pi}{c\left(h\left(2.0895+\ln\left[1.444+\frac{t_w}{h}\right]\right)+1.5t_w\right)}, & \text{otherwise} \end{cases}, \tag{4}$$

and h is the substrate thickness.

The calculated values of f_0, C, and L define the dimensions considered in the EM simulations. It is also a starting point where the designer must proceed a fine-tune to achieve a desired resonance center frequency. Also, a second method one can use is scaling based on a predesigned structure. The rejection band moves to lower frequencies as the structure is scaled up and vice-versa. Furthermore, the attenuation further improves if the PBG structure forms a lattice. As the designer works at higher frequency ranges, UCPBG achieves smaller geometric sizes, improving its applicability. In this case, metal slots are etched in the ground plane connected by narrow lines to form a distributed LC network [17]. This method fits well for standard PCB designs because it is planar, countering the need for vias or unique materials, saving costs. Concerning the hexagonal geometry, the reduction in the PCB-occupied area is the most attractive design feature. It can keep the same gap perimeter on the side of a square while occupying 65% of the area.

The current work investigates six variations of HCPBG geometries. The desired f_0 is achieved by simply changing the number of traces, introducing different gap lengths, rotation of PBG structure relative to the microstrip line, and geometric scaling. Table 1 describes and illustrates the variations proposed for simulations and measurements. Each line of the table constitutes the structure case, the number of traces, rotation, and the PCB bottom layer PBG structure relative to the microstrip line represented as dashed contours. The cases of Table 1 (a) and (b) explores the asymmetry in the HCPBG structure by suppressing some of the traces, which changes the LC elements from the basic geometry presented in Figure 2b.

The capacitance of the gap and the trace inductance closer to the point where the transmission line crosses the PBG have a more substantial influence on the resonance frequency. Adding asymmetry changes the values of the LC and the influence over the resonances.

The cases (a) and (b) are the three traces and show asymmetries for both gaps and traces close to the microstrip line. For cases (c) and (d), we can notice that traces and gaps are symmetrically apart from the microstrip line, and in case (e), there are two traces parallel to the microstrip line due to the rotation of 30°, for which it is expected to have similar results to the UCPBG. Finally, the case (f) is the fundamental geometry presented in Figure 2b.

Table 1. Six variations of HCPBG geometries considering rotations and suppressed traces.

Label	Traces	Rotation	Structure
(a)	3	0°	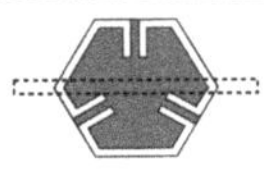
(b)	3	30°	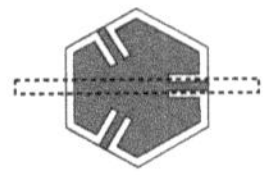
(c)	4	0°	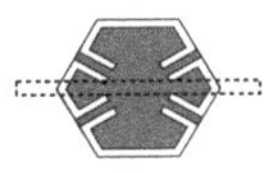
(d)	4	90°	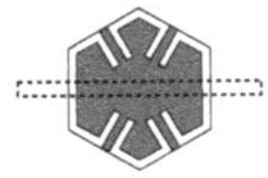
(e)	4	30°	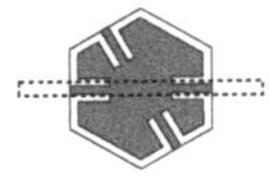
(f)	6	0°	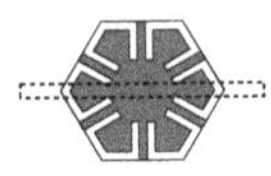

3. Simulation of HCPBG

The simulation setup consists of a reference plane PCB (without PBG on the ground layer), a UCPBG, and six HCPBG combinations shown in Table 1. Six out of seven simulations on HCPBG structures have dimensions defined in Table 2 case (a). The seventh simulation has dimensions defined in Table 2 case (b), where g_2 is the only different parameter. We calculated the PCB's frequency profile using CST Studio® [28], an electromagnetic field simulation software.

In the CST Studio simulation, we set the time domain solver, which employs finite integration technique (FIT) [29]. The time domain solver has the ability to handle large and complex structures, and it allows for memory efficient computation. We set the accuracy to −40 dB and hexahedral mesh type. For the mesh properties, we defined 12 cells per wavelength near/far from model and 35 cells of fraction minimum cell near model.

The PCB is an FR-4 two-layer of 90 mm length, 60 mm width, and 1.6 mm thickness. The microstrip line is 50 Ω, 3 mm width and 90 mm length on the top layer. The ground plane is placed on the bottom side of the PCB, where the HCPBG and UCPBG structures are located, as shown in Figure 1. The simulations consider the dimensions of Table 2 case (a) where $L = 3.26$ nH, $C = 1.38$ pF and $f_0 = 2.37$ GHz when using Equations (2) and (3). The nomenclature to name the HCPBG structures is defined as HCPBG-xx-yy, where "xx" represents the number of traces and "yy" is the orientation relative to the microstrip line.

Table 2. Dimensions used for UCPBG and HCPBG designs.

Case	a	g_1	g_2	t_w	t_l
(a)	28.5 mm	1.5 mm	1.5 mm	7.5 mm	1.5 mm
	$0.23\lambda_{2.37\text{ GHz}}$	$0.012\lambda_{2.37\text{ GHz}}$	$0.012\lambda_{2.37\text{ GHz}}$	$0.06\lambda_{2.37\text{ GHz}}$	$0.012\lambda_{2.37\text{ GHz}}$
(b)	28.5 mm	1.5 mm	0.5 mm	7.5 mm	1.5 mm
	$0.23\lambda_{2.37\text{ GHz}}$	$0.012\lambda_{2.37\text{ GHz}}$	$0.04\lambda_{2.37\text{ GHz}}$	$0.06\lambda_{2.37\text{ GHz}}$	$0.012\lambda_{2.37\text{ GHz}}$

The most significant characteristic of HCPBG unit cell is that multiple resonance profiles emerge due to rotation and trace suppression. Figure 3 shows the transmission loss profile for UCPBG and HCPBG Table 1 case (e). Both have similar results up to 3.3 GHz, with the first resonance at 2.36 GHz and similar bandwidth (BW); BW = 136 MHz at −10 dB. This case presents a second strong resonance at 4.13 GHz with BW = 100 MHz and the third one at 5.38 GHz, but with an attenuation lower than the −10 dB. This third resonance coincides with the second resonance of the UCPBG structure but with lower attenuation.

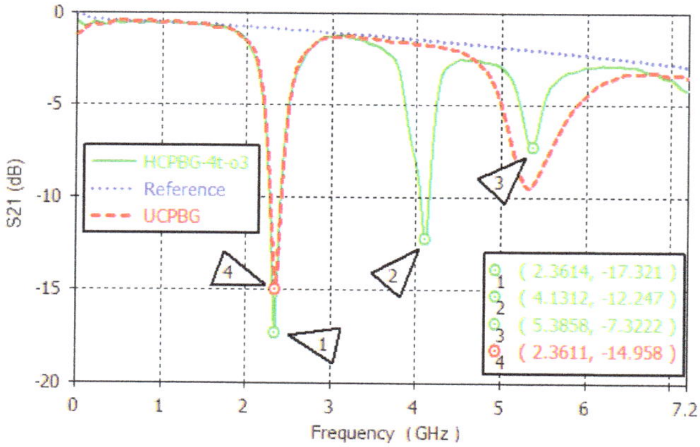

Figure 3. Simulation—transmission loss for UCPBG and HCPBG-4t-o3.

Figures 4 highlights the effect of HCPBG rotation for Table 1 (c) and (d) cases. HCPBG and UCPBG present a similar filtering profile observed in the previous simulation. Notice that the main differences are due to the rotation of HCPBG-4t-o1 and HCPBG-4t-o2 (Table 1 case (c) and (d), respectively). The HCPBG-4t-o1 has two strong attenuated bands, one at 2.17 GHz (BW = 167 MHz) and the second at 3.61 GHz, with a wide bandwidth (BW = 982 MHz). In HCPBG-4t-o2, the first attenuation band has an extensive bandwidth at 2.25 GHz (BW = 1.4 GHz), and the second resonance locates at 4.14 GHz (BW = 260 MHz).

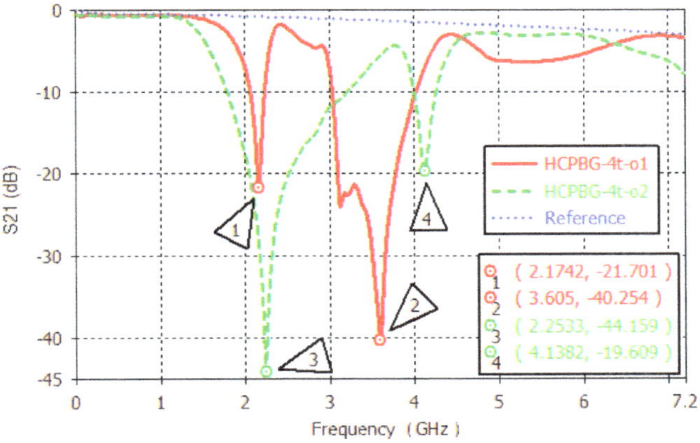

Figure 4. Simulation—transmission loss for HCPBG with 4 traces—orientations o1 and o2.

We also simulated HCPBG with three traces and six traces, keeping the same dimensions to verify the attenuation band profile's behavior. Figure 5 shows the transmission loss for the HCPBG with 3 traces and two different orientations, as indicated in Table 1 by cases (a) and (b). Both have similar filtering profiles up to 4.5 GHz, with a first attenuation around 2 GHz, and with BW of 645 MHz for HCPBG-3t-01 and BW of 380 MHz HCPBG-3t-o2. The second attenuation band is close to 3 GHz for both. The main difference is that HCPBG-3t-o1 produces a third strong wideband attenuation at 5.75 MHz (BW = 1670 MHz). The LC network seen by the microstrip line depends on the HCPBG orientation, and it seems to have more impact when the HCPBG traces are more aligned with the microstrip line.

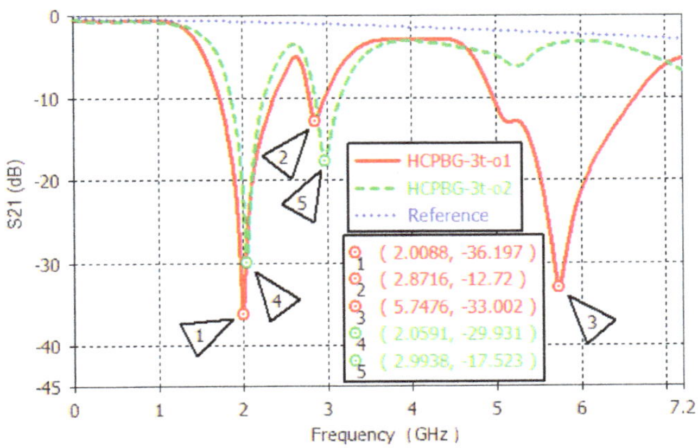

Figure 5. Simulation—transmission loss for HCPBG with 3 traces—orientations o1 and o2.

Figure 6 shows the transmission loss simulation results for the 6-trace HCPBG, Table 1 case (f), with outer gaps of 1.5 and 0.5 mm. A smaller gap results in a greater capacitance, leading to lower rejection band frequencies. The HCPBG-6t with a gap of 1.5 mm has a significant bandwidth response if we use a criterion of −5.8 dB (BW = 3.16 GHz). This structure could be helpful in the suppression of broadband noise.

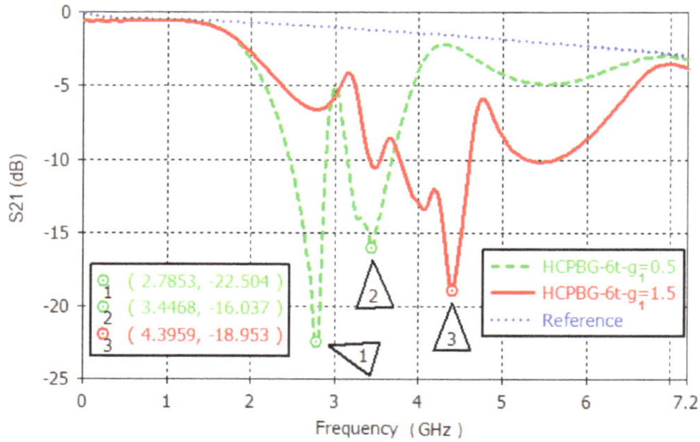

Figure 6. Simulation—transmission loss for HCPBG with 6 traces—gap distances 1.5 and 0.5 mm.

4. Measurements of HCPBG

We also verified the analytical guidelines and simulation results performing measurements in an actual PCB. An LPKF machine milled the 2-layer FR-4 samples of UCPBG and HCPBG structures. Figure 7 shows the bottom side of two fabricated samples, UCPBG (to the left) and the HCPBG-4t-o3 (to the right). The HCPBG-3t-o1, HCPBG-3t-o2, HCPBG-6t (g = 1.5 mm), and the reference board (ground only) were also fabricated. The sample dimensions and substrates are the same used for the simulations.

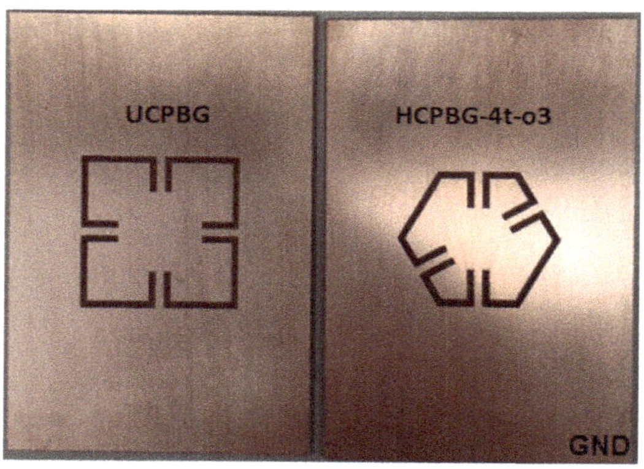

Figure 7. Fabricated UCPBG (**left**) and HCPBG with 4 traces, orientation 3 (**right**)-PCB bottom view.

The measurement setup, shown in Figure 8, is composed of the Keysight Field Fox RF Analyzer N9914B, 2 RF cables, N-SMA adapters, and the samples. The analyzer is set to network analyzer mode and calibrated from 30 kHz to 6.5 GHz.

The measured logarithmic magnitude of S21 parameter for the UCPBG and HCPBG-4t-o3 can be seen in Figure 9. When comparing the results in Figure 3, we can observe that the simulation curves' profiles are very close to the measurement results. However, the resonance frequencies showed a slight discrepancy, with the UCPBG's case being the more evident one. This discrepancy is probably due to fabrication process variations, such as gap width, depth, and substrate dielectric constant variation. Nonetheless, the simulation results showed good accuracy that can be observed in all measurements. The UCPBG's measured first resonance frequency is at 2.18 GHz (BW = 160 MHz), and the second one at 5.4 GHz. For the HCPBG-4t-o3, the resonance frequencies are 2.4 MHz (BW = 160 MHz), 4.26 GHz (BW = 130 MHz), and 5.56 GHz. The bandwidth and maximum attenuation are also very similar in both measurement and simulation. The reference line represents the measured S21 for the transmission line with a solid ground plane (no PBG).

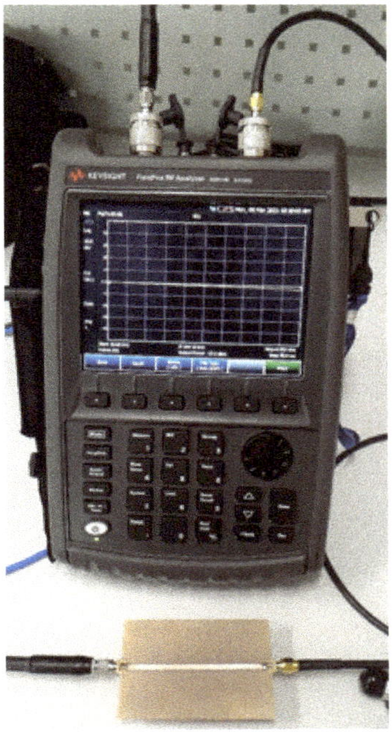

Figure 8. Measurement Setup for Transmission Loss.

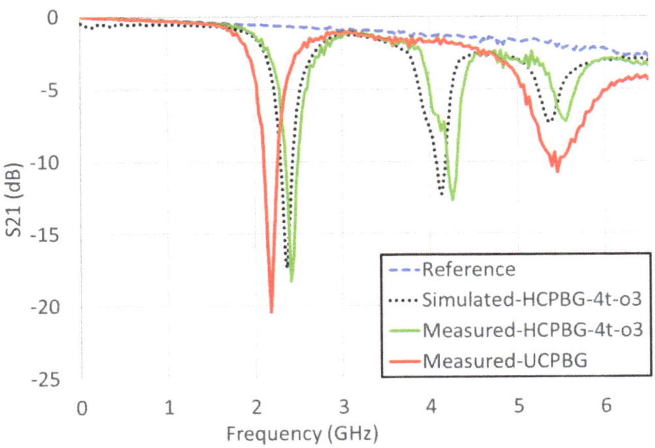

Figure 9. Measurement—transmission loss for HCPBG with 4 traces, orientation 3 vs. UCPBG.

The graphics of Figure 10 show the measurement results for the three trace geometries: HCPBG-3t-o1 and HCPBG-3t-o2. The simulation results are in good accordance with the measurements concerning curve profile, rejection band center frequency, and bandwidth. Similarly to the simulation, the HCPBG-3t-o1 first attenuation band is at 2.04 GHz (BW = 650 MHz), and for HCPBG-3t-o2, it is at 2.08 GHz (BW = 390 MHz). The second resonance appears at 2.92 GHz (BW = 260 MHz) and 3.05 GHz (BW = 292 MHz) for orientations 1 and 2, respectively. Also, as predicted by simulation, a third wide resonance occurs at 6.01 GHz (BW = 1.4 GHz) for the case of HCPBG-3t-o1.

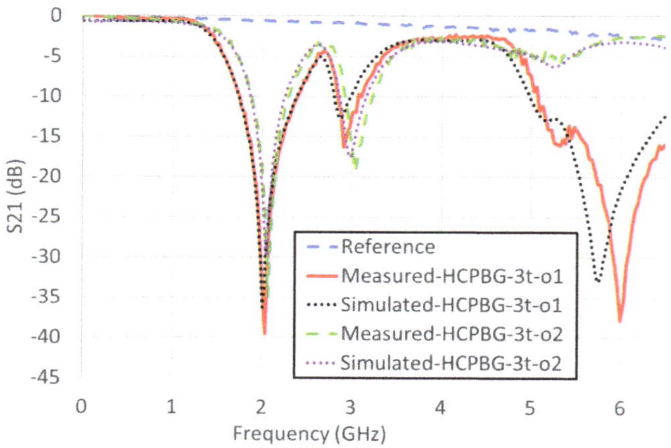

Figure 10. Measurement—transmission loss for HCPBG with 3 traces, orientations 1 and 2.

Finally, Figure 11 shows the transmission loss measurement results for the six trace HCPBG with a gap of 1.5 mm. Although it shows a significant bandwidth response and a similar profile, when compared to that of the simulated model, the bandwidth is reduced when we use −5.8 dB criteria. To have similar bandwidth, the criteria would need to be around −5 dB. The more substantial attenuation is at 4.55 GHz (−18.8 dB).

The simulation results show an error of less than 2% in the resonance frequencies below 4 GHz, while for values above 4 GHz, the error is less than 4%. The discrepancy between the measurements and simulation results for higher frequencies could be due to a mismatch between the dielectric characteristics of the PCB and the simulation model, over frequency. The variations in the fabrication process of the board would also impact the structure geometry; for example, changing the gap capacitance and displacing the resonance frequency.

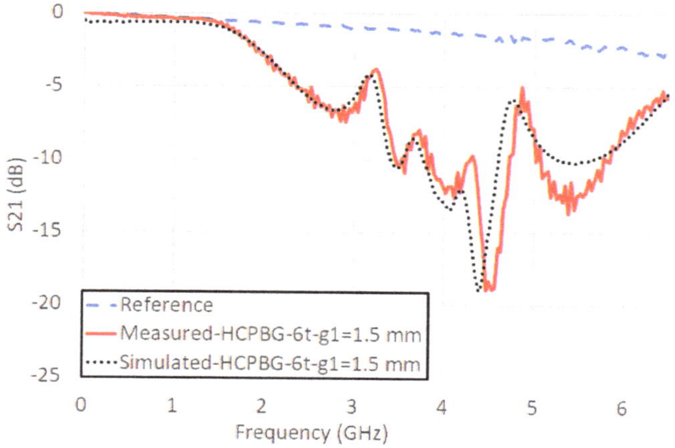

Figure 11. Measurement—transmission loss for HCPBG with 6 traces—gap distances 1.5 mm.

5. Design of Reconfigurable HCPBG

First, it is essential to remind that the central concept of HCPBG concerns the model of two metal plates separated by gaps of etched copper. Also, the traces of each hexagon face and gaps define the features of an equivalent LC network. Then, it is possible to tune the resonance center frequency of the filter by actively controlling the parameters L and C. The simplest method to achieve frequency tuning is to switch on/off one or more traces that result in inductance changes. For this purpose, it is possible to use FET transistors or micro-electro-mechanical systems (MEMS) switches. The caveat of this approach is the additional capacitance of the FET transistor or MEMS while in an off state, making the implementation more complex.

On the other hand, it is also possible to change the LC network using a variable capacitor at the gaps close to the transmission line. The technique allows actively reconfigure the first resonance frequency adding parallel capacitance to the overall intrinsic gap capacitance. In this case, a varactor diode connected to a bias tee circuit with a port connected to a DC voltage supply allows controlling the capacitance. The varactor diode method is less complex than active inductor circuits or RF switches and provides fine-tuning of the first resonance frequency. Hence, the cost of implementing switches or the complexity of polarizing FET transistors for each trace of the HCPBG would be higher than using two varactor diodes to change the capacitance.

The selected HCPBG model for the reconfiguration study is the 6-trace type, and it can be observed in Table 1 case (f). The main dimensions used to create the reconfigurable HCPBG structure can be seen in Table 3.

Table 3. Reconfigurable HCPBG dimensions.

a	g_1	g_2	t_l	t_w
28.5 mm	0.5 mm	0.5 mm	7.5 mm	1.5 mm

The model shown in Figure 12 was simulated in the CST studio software. It consists of a 50 Ω microstrip line on a 2-layer FR4 PCB, with a dielectric constant of 4.3, length of 90 mm, a width of 60 mm, and thickness of 1.6 mm. The ground plane is placed on the bottom side of the PCB, where the HCPBG structure is also etched. All pads, traces, and vias are designed to emulate the fabricated board.

On the bottom layer, we have the varactor diode ($D_{\_varicap}$) and the DC block capacitor ($C_{\_dcblock}$). The inductor ($L_{\_choke}$) to isolate the DC from the AC part of the circuit is placed on the top layer. A metal via connects the DC power supply traces from the top layer to the varactor on the bottom layer. At the end of the DC power supply trace, a capacitor ($C_{\_supply}$) is connected to the ground to represent the power supply's capacitive coupling to the common ground.

A simplified series RLC varactor model is used in the simulation. It is composed of a capacitor (C_T) in series with the parasitic inductance (L_S) and resistance (R_S). According to the datasheet of SMV1247 from Skyworks, L_S is 0.7 nH for the SC-79 package, and R_S is dependent on the applied reverse voltage (V_R), having a value that ranges from 2.5 to 9 Ω. The applied reverse voltage controls the varactor's capacitance (C_T). The rationale behind the choice of the SM1247 is that its capacitance range is in the same order as the gap capacitance of the HCPBG, giving a good dynamic range for frequency shifting. Table 4 shows selected C_T versus V_R used for the simulations.

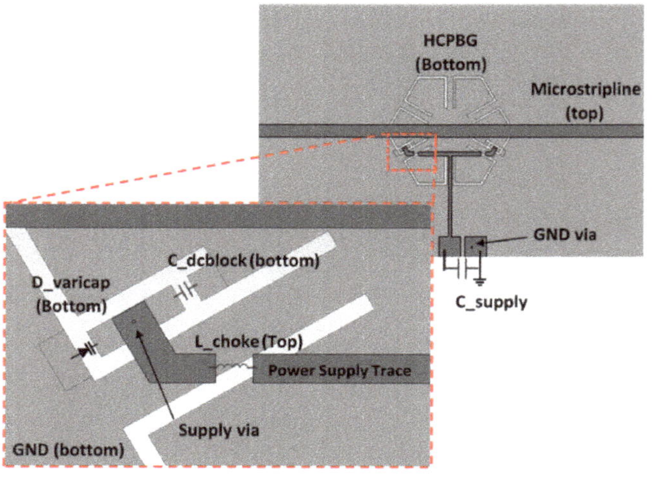

Figure 12. Reconfigurable HCPBG—detailed simulation design view.

Table 4. Reverse voltage (V_R) versus capacitance (C_T) of SMV1247 varactor diode.

V_R	7.5 V	4.0 V	3.5 V	3 V	2.5 V	2.0 V	0.0 V
C_T	0.64 pF	0.77 pF	0.83 pF	0.95 pF	1.22 pF	1.88 pF	8.86 pF

The representation of the varactor's circuit model (C_T, R_S, L_S) and bias circuit electrical connections can be seen in Figure 13. The varactor's anode is connected to a common ground plane (PCB GND) and the cathode to the DC block capacitor (C_{dc_blk}), which is also connected to the HCPBG's inner GND. This configuration allows DC bias isolation from the RF ground, for the C_T works as a second DC block element. Completing the bias tee, the inductor ($L_{\_choke}$) isolates the RF signal from the DC power supply line. The DC power supply coupling capacitance ($C_{\_supply}$) connects the DC line to the PCB GND.

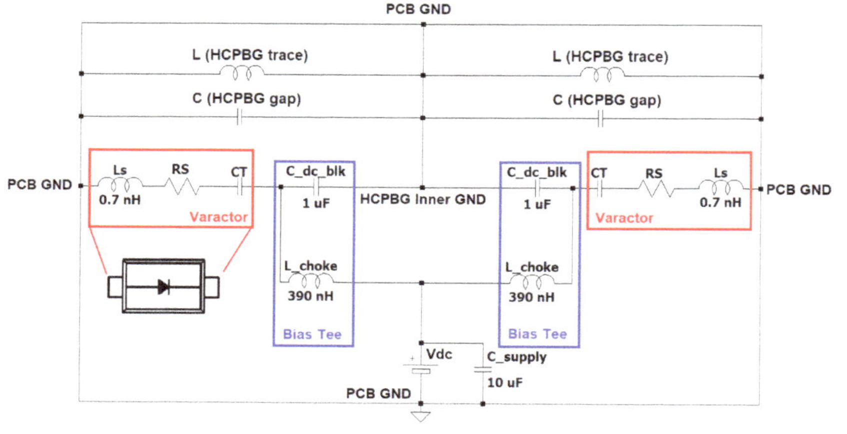

Figure 13. Reconfigurable HCPBG—varactor and bias tee schematic.

The discrete component values used in the simulation can be seen in Table 5.

Table 5. Simulation—discrete component values.

C_T	L_S	R_S	$C_{\_dcblock}$	$L_{\_choke}$	$C_{\_supply}$
0.64–8.86 pF	0.7 nH	2.5–9 Ω	1 μF	0.39 μH	10 μF

To validate the simulation model, we fabricated the testing sample using an LPKF milling machine. In Figure 14a, is the top view of the fabricated sample, showing the microstrip line, SMA connectors, choke inductors, and DC power supply traces and pads. The bottom side of the sample is shown in Figure 14b, containing the GND plane, HCPBG, varactor diodes, and DC block capacitors.

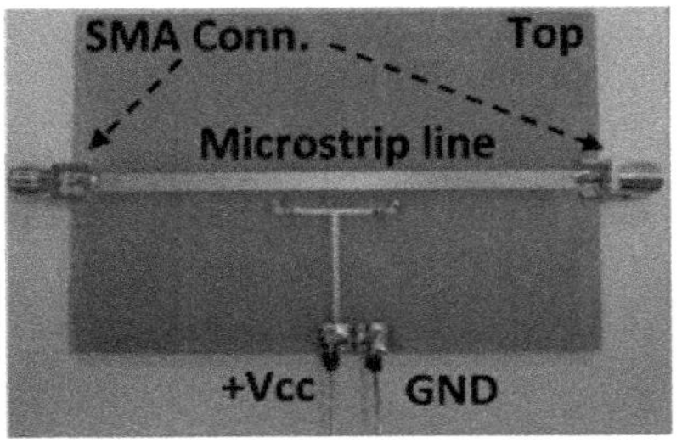

(a)

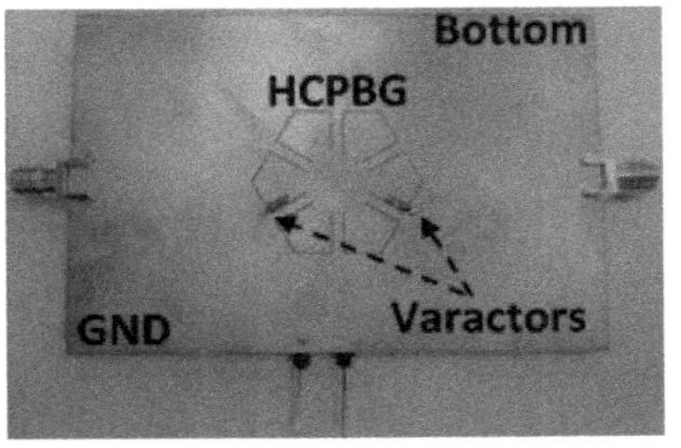

(b)

Figure 14. Reconfigurable HCPBG-6t-fabricated sample. (**a**) Top view; (**b**) bottom view.

The filtering characteristic of the reconfigurable HCPBG can be observed by the simulated and measured transmission loss (S21) in Figures 15 and 16, respectively. Tables 6 and 7 show the simulated and measured resonance frequencies and bandwidths for different V_R or C_T values.

Both results show similar curve profiles, and reasonable approximation with respect to resonance center frequency and bandwidth. For instance, considering C_T of 0.64 pF (V_R = 7.5 V), the simulated first resonance frequency f_1 is 2.49 GHz (BW_1 = 331 MHz) and the measured f_1 is 2.31 GHz (BW_1 = 260 MHz).

Starting the analysis with the simulation results, as we apply the lowest C_T value (0.64 pF) two main resonances occur, the first one at 2.49 GHz (BW_1 = 331 MHz) and the second one at 4.63 GHz (BW_1 = 647 MHz). As expected, when we increase C_T, the resonance frequencies are shifted to lower values. For example, if we take C_T values of 0.83 pF and 0.95 pF, f_1 is displaced by 100 MHz to 2.32 GHz (BW_1 = 244 MHz) and 2.22 GHz (BW_1 = 209 MHz), respectively. The same behavior is observed for the measurement results; for example, for V_R of 3.5 V (0.83 pF) and 3.0 V (0.95 pF), the f_1 is displaced by 70 MHz, going from 2.15 GHz (BW_1 = 160 MHz) to 2.08 GHz (BW_1 = 160 MHz).

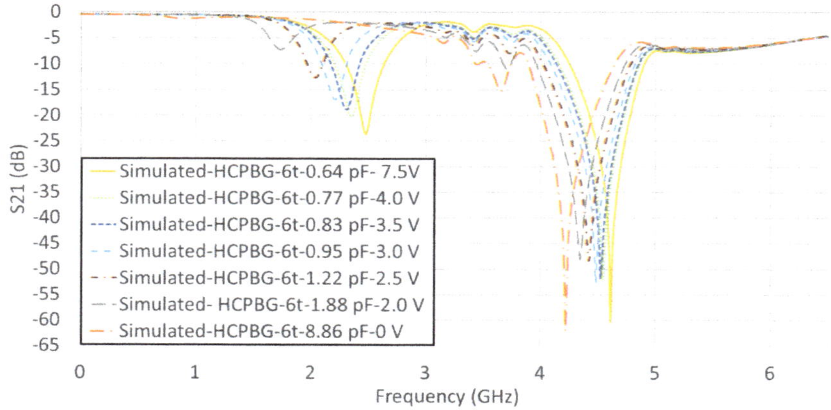

Figure 15. Simulated—transmission loss for voltage values: 7.5, 4.0, 3.5, 3.0, 2.5, 2.0, 0.0 Volts.

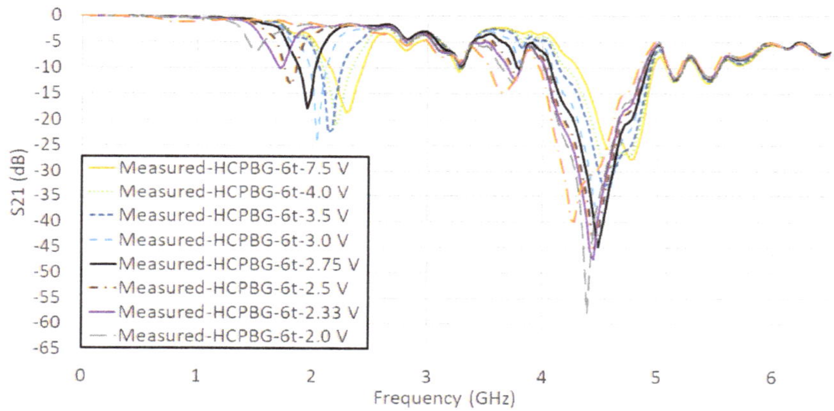

Figure 16. Measured—transmission loss for voltage values: 7.5, 4.0, 3.5, 3.0, 2.75, 2.5, 2.33, 2.0, 0.0 Volts.

Table 6. Simulation—stop band center frequency and bandwidth (BW).

V_R (pF)	C_T (pF)	f_1 (GHz)	BW_1 (MHz)	f_2 (GHz)	BW_2 (MHz)
7.5	0.64	2.49	331	4.63	647
4.0	0.77	2.36	266	4.56	690
3.5	0.83	2.32	244	4.53	712
3.0	0.95	2.22	209	4.50	733
2.5	1.22	2.02	137	4.42	755
2.0	1.88	1.73	(−7.29 dB)	4.34	762
0.0	8.86	0.98	(−1.35 dB)	4.23	741

Table 7. Measurement—stop band center frequency and bandwidth (BW).

V_R (pF)	C_T (pF)	f_1 (GHz)	BW_1 (MHz)	f_2 (GHz)	BW_2 (MHz)
7.5	0.64	2.31	260	4.78	650
4.0	0.77	2.21	230	4.58	710
3.5	0.83	2.15	160	4.55	750
3.0	0.95	2.08	160	4.52	810
2.75	-	1.95	130	4.49	780
2.50	1.22	1.82	97	4.45	810
2.33	-	1.72	(−10.19 dB)	4.45	840
2.0	1.88	1.5	(−6.97 dB)	4.39	810
0.0	8.86	0.85	(−1.26 dB)	4.29	840

The difference between the simulated and measured resonance frequencies is dependent on the varactor's bias voltage or the selected C_T. Considering C_T values of 1.22 pF and above, the first resonance frequency showed an error of 15%. On the other hand, for capacitance values of 0.95 pF and bellow, the difference is less than 8%. According to the datasheet, the varactor SMV1247 can show 7% to 11% variation in the capacitance, depending on the applied voltage. Added to that, we have the fabrication process and substrate dielectric variations. For the second resonance, around 4.5 GHz, for which the varactor does not play a strong influence, the error is only 3%.

Another observed characteristic is the bandwidth and maximum attenuation reduction, as the resonance is shifted to lower frequencies. This reduction is strongly influenced by the varactor's R_S. For example, considering C_T of 0.77 pF, then R_S is 3.9 Ω and simulated S21 equals to −20 dB @ 2.36 GHz. For C_T of 1.22 pF, then R_S = 6.3 Ω and simulated S21 equals to −12.6 dB @ 2.02 GHz. Due to R_S effect, the −10 dB criteria can not be achieved for 1.88 pF (S21 = −7.29 dB @ 1.73 GHz) and 8.86 pF (S21 = −1.35 dB @ 0.98 GHz), limiting the dynamic range of the reconfigurable filter. The measurement results confirm this characteristic. When we set V_R to 2.0 V, or equivalently, C_T to 1.88 pF, the filter attenuation stays above the −10 dB criteria (S21 = −6.97 dB @ 1.5 GHz). If we remove R_S from the simulation, the bandwidth is also reduced as the frequency shifts, however the entire range of C_T achieves the −10 dB attenuation criteria.

C_T has a stronger influence over the first resonance frequency. For example, in Table 6, if we observe the resonance frequencies at 0.64 pF and 8.86 pF, f_1 shifts 1.51 GHz, however, the second resonance (f_2) shifts only 400 MHz.

For a −10 dB criteria, the measurements show a filtering capability that ranges from 1.72 GHz (V_R = 2.33 V) to 2.44 GHz (V_R = 7.5 V, upper band of f_1 = 2.31 GHz), that is 720 MHz bandwidth coverage.

6. Conclusions

In this paper, we presented a new planar PBG geometry based on the UCPBG: the HCPBG. It occupies less PCB area on the reference plane allowing a more compact usage of the space area. The simulations showed good accuracy when compared to that of the measurements. We showed through simulation and measurement that HCPBG unit cell structure can produce similar characteristics to the UCPBG, but it also adds additional resonance bands of interest. Furthermore, the rejection band is reconfigurable depending on the orientation and number of traces used, which results in more flexibility on filter designs for interfering signals. We also presented a reconfigurable HCPBG single-cell structure that can be controlled electronically. It allows for changing the rejection band profile by applying DC voltage at the control port. The control circuit, composed of two varactor diodes and a bias tee circuit, allows changing the gap capacitance of the HCPBG structure. The concept was demonstrated by EM simulation, including the varactor model and bias tee components. The measurement results point out that the resonance center frequency changes proportionally to the varactor's capacitance, thereby agreeing with the results observed in simulations. The simulated model showed less than 4% error in the resonance center frequency. For higher frequencies, the difference is stronger, probably due to variations in the fabrication process and a model mismatch of the PCB dielectric characteristics. In the case of reconfigurable HCPBG, we observed that the varactor's capacitance plays a strong role in the difference between the simulated and measured S parameters. The first simulated resonances showed 15% differences with respect to the measured ones, however, the second resonance has a maximum of 3% variation.

Author Contributions: Conceptualization, A.T.d.A.S.; methodology, validation, formal analysis, investigation, A.T.d.A.S., C.F.D., E.R.d.L. and G.F.; resources, E.R.d.L. and L.M.d.A.; data curation, writing—original draft preparation, A.T.d.A.S.; writing—review and editing, A.T.d.A.S., C.F.D., E.R.d.L., G.F. and L.M.d.A.; visualization, supervision, project administration, funding acquisition, E.R.d.L. and L.M.d.A. All authors have read and agreed to the published version of the manuscript.

Funding: This research received no external funding.

Acknowledgments: The authors are grateful to R&D Project PD-07130-0047, funded by Transmissora Aliança de Energia Elétrica SA (TAESA), with resources from ANEEL R&D ProgramProgram and São Paulo Research Foundation (FAPESP) grant #15/24494-8.

Conflicts of Interest: The authors declare no conflict of interest.

Abbreviations

The following abbreviations are used in this manuscript:

HCPBG	Honey Comb Photonic Band Gap
UCPBG	Uniplanar Compact Photonic Band Gap
PBG	Photonic Band Gap
EM	Electromagnetic
FR1	Frequency Range 1
PCB	Printed Circuit Board
EBG	Electromagnetic Band Gap
EMC	Electromagnetic Compatibility
RIS	Reconfigurable Intelligent Surface
MEMS	Microelectromechanical Systems
BW	Bandwidth

References

1. Herbertz, K. Design and Characterisation of Electromagnetic Bandgap Filters. Ph.D. Thesis, Imperial College London, London, UK, 2010.
2. Kim, W.; Lee, B. Modelling and design of 2D UC-PBG structure using transmission line theory. *IEEE Antennas Propag. Soc. Int. Symp.* **2002**, *3*, 780–783.
3. Shahparnia, S.; Ramahi, O.M. Electromagnetic interference (EMI) reduction from printed circuit boards (PCB) using electromagnetic bandgap structures. *IEEE Trans. Electromagn. Compat.* **2004**, *46*, 580–587. [CrossRef]
4. Qin, J.; Ramahi, O. Novel wideband planar electromagnetic bandgap structures for noise suppression. In Proceedings of the 2006 IEEE Workshop on Signal Propagation on Interconnects, Berlin, Germany, 9–12 May 2006; pp. 83–86.
5. Wu, T.L.; Lin, Y.H.; Chen, S.T. A novel power planes with low radiation and broadband suppression of ground bounce noise using photonic bandgap structures. *IEEE Microw. Wirel. Compon. Lett.* **2004**, *14*, 337–339.
6. Elamaran, B.; Chio, I.M.; Chen, L.Y.; Chiao, J.C. A beam-steerer using reconfigurable PBG ground plane. In Proceedings of the 2000 IEEE MTT-S International Microwave Symposium Digest (Cat. No. 00CH37017), Boston, MA, USA, 11–16 June 2000; Volume 2, pp. 835–838.
7. Alibakhshikenari, M.; Virdee, B.S.; Shukla, P.; See, C.H.; Abd-Alhameed, R.A.; Falcone, F.; Quazzane, K.; Limiti, E. Isolation enhancement of densely packed array antennas with periodic MTM-photonic bandgap for SAR and MIMO systems. *IET Microw. Antennas Propag.* **2019**, *14*, 183–188. [CrossRef]
8. Saini, J.; Garg, M.K. PBG Structured Compact Antenna with Switching Capability in Lower and Upper Bands of 5G. *Prog. Electromagn. Res.* **2020**, *94*, 19–29. [CrossRef]
9. Mollah, M.; Karmakar, N. Planar PBG structures and their applications to antennas. In Proceedings of the 2001 IEEE Antennas and Propagation Society International Symposium, Boston, MA, USA, 8–13 July 2001; Volume 2, pp. 494–497.

10. Jiménez-Guzmán, G.Á.; Tirado-Mendez, J.A.; Rangel-Merino, A.; Vasquez-Toledo, L.A.; Marcelin-Jimenez, R. Improving the performance of a patch antenna array by using photonic bandgap structures at X-band. *J. Electromagn. Waves Appl.* **2020**, *34*, 2130–2146. [CrossRef]
11. Mantash, M.; Kesavan, A.; Denidni, T.A. Beam-tilting endfire antenna using a single-layer FSS for 5G communication networks. *IEEE Antennas Wirel. Propag. Lett.* **2017**, *17*, 29–33. [CrossRef]
12. Tsilipakos, O.; Tasolamprou, A.C.; Pitilakis, A.; Liu, F.; Wang, X.; Mirmoosa, M.S.; Tzarouchis, D.C.; Abadal, S.; Taghvaee, H.; Liaskos, C.; et al. Toward Intelligent Metasurfaces: The Progress from Globally Tunable Metasurfaces to Software-Defined Metasurfaces with an Embedded Network of Controllers. *Adv. Opt. Mater.* **2020**, *8*, 2000783. [CrossRef]
13. Björnson, E.; Özdogan, Ö.; Larsson, E.G. Reconfigurable intelligent surfaces: Three myths and two critical questions. *IEEE Commun. Mag.* **2020**, *58*, 90–96. [CrossRef]
14. Liaskos, C.; Nie, S.; Tsioliaridou, A.; Pitsillides, A.; Ioannidis, S.; Akyildiz, I. A new wireless communication paradigm through software-controlled metasurfaces. *IEEE Commun. Mag.* **2018**, *56*, 162–169. [CrossRef]
15. Karbassian, M.; Ghafouri-Shiraz, H. Effect of shape of patterns on the performance of microstrip photonic band-gap filters. *Microw. Opt. Technol. Lett.* **2006**, *48*, 1007–1011. [CrossRef]
16. Yang, F.R.; Coccioli, R.; Qian, Y.; Itoh, T. Planar PBG structures: Basic properties and applications. *IEICE Trans. Electron.* **2000**, *83*, 687–696.
17. Yang, F.R.; Ma, K.P.; Qian, Y.; Itoh, T. A uniplanar compact photonic-bandgap (UC-PBG) structure and its applications for microwave circuit. *IEEE Trans. Microw. Theory Tech.* **1999**, *47*, 1509–1514. [CrossRef]
18. Liang, B.; Sanz-Izquierdo, B.; Parker, E.A.; Batchelor, J.C. A frequency and polarization reconfigurable circularly polarized antenna using active EBG structure for satellite navigation. *IEEE Trans. Antennas Propag.* **2014**, *63*, 33–40. [CrossRef]
19. Dewan, R.; Abd Rahim, M.K.; Hamid, M.R.; Himdi, M.; Majid, H.B.A.; Samsuri, N.A. HIS-EBG unit cells for pattern and frequency reconfigurable dual band array antenna. *Prog. Electromagn. Res.* **2018**, *76*, 123–132. [CrossRef]
20. Hill, M.J.; Ziolkowski, R.W.; Papapolymerou, J. A high-Q reconfigurable planar EBG cavity resonator. *IEEE Microw. Wirel. Compon. Lett.* **2001**, *11*, 255–257. [CrossRef]
21. Zhou, W.; Mackie, D.M.; Taysing-Lara, M.; Dang, G.; Newman, P.G.; Svensson, S. Novel reconfigurable semiconductor photonic crystal-MEMS device. *Solid-State Electron.* **2006**, *50*, 908–913. [CrossRef]
22. Fourn, E.; Pothier, A.; Champeaux, C.; Tristant, P.; Catherinot, A.; Blondy, P.; Tanné, G.; Rius, E.; Person, C.; Huret, F. MEMS switchable interdigital coplanar filter. *IEEE Trans. Microw. Theory Tech.* **2003**, *51*, 320–324. [CrossRef]
23. Karim, M.; Liu, A.; Alphones, A.; Yu, A. A novel reconfigurable filter using periodic structures. In Proceedings of the 2006 IEEE MTT-S International Microwave Symposium Digest, San Francisco, CA, USA, 11–16 June 2006; pp. 943–946.
24. Karim, M.; Guo, Y.X.; Chen, Z.; Ong, L. Miniaturized reconfigurable and switchable filter from UWB to 2.4 GHz WLAN using PIN diodes. In Proceedings of the 2009 IEEE MTT-S International Microwave Symposium Digest, Boston, MA, USA, 7–12 June 2009; pp. 509–512.
25. Kurra, L.; Abegaonkar, M.P.; Basu, A.; Koul, S.K. Bandwidth reconfigurable bandstop filter using planar EBG structure. In Proceedings of the IEEE MTT-S International Microwave and RF Conference, New Delhi, India, 14–16 December 2013; pp. 1–3.
26. Paul, C.R. *Analysis of Multiconductor Transmission Lines*; John Wiley & Sons: Hoboken, NJ, USA, 2007.
27. Paul, C.R. *Inductance: Loop and Partial*; John Wiley & Sons: Hoboken, NJ, USA, 2011.
28. Systemes, D. CST Studio Suite 2021.
29. Hirtenfelder, F. Effective antenna simulations using CST MICROWAVE STUDIO®. In Proceedings of the 2007 2nd International ITG Conference on Antennas, Munich, Germany, 28–30 March 2007; p. 239. [CrossRef]

electronics

Article

Coordinate Transformations-Based Antenna Elements Embedded in a Metamaterial Shell with Scanning Capabilities

Dipankar Mitra [1,*], Sukrith Dev [2], Monica S. Allen [2], Jeffery W. Allen [2] and Benjamin D. Braaten [1,*]

1 Department of Electrical and Computer Engineering, North Dakota State University, Fargo, ND 58105, USA
2 Air Force Research Laboratory, Munitions Directorate, Eglin Air Force Base, FL 32542, USA; sukrith.dev.1@us.af.mil (S.D.); monica.allen.3@us.af.mil (M.S.A.); jeffery.allen.12@us.af.mil (J.W.A.)
* Correspondence: dipankar.mitra@ndsu.edu (D.M.); benjamin.braaten@ndsu.edu (B.D.B.)

Abstract: In this work transformation electromagnetics/optics (TE/TO) were employed to realize a non-homogeneous, anisotropic material-embedded beam-steerer using both a single antenna element and an antenna array without phase control circuitry. Initially, through theory and validation with numerical simulations it is shown that beam-steering can be achieved in an arbitrary direction by enclosing a single antenna element within the transformation media. Then, this was followed by an array with fixed voltages and equal phases enclosed by transformation media. This enclosed array was scanned, and the proposed theory was validated through numerical simulations. Furthermore, through full-wave simulations it was shown that a horizontal dipole antenna embedded in a metamaterial can be designed such that the horizontal dipole performs identically to a vertical dipole in free-space. Similarly, it was also shown that a material-embedded horizontal dipole array can perform as a vertical dipole array in free-space, all without the need of a phase shifter network. These methods have applications in scanning for wireless communications, radar, beam-forming, and steering.

Keywords: transformation electromagnetics/optics; coordinate transformations; meta-material; beam-steering

Citation: Mitra, D.; Dev, S.; Allen, M.S.; Allen, J.W.; Braaten, B.D. Coordinate Transformations-Based Antenna Elements Embedded in a Metamaterial Shell with Scanning Capabilities. *Electronics* **2021**, *10*, 1081. https://doi.org/10.3390/electronics10091081

Academic Editors: Naser Ojaroudi Parchin, Mohammad Ojaroudi and Raed A. Abd-Alhameed

Received: 9 April 2021
Accepted: 1 May 2021
Published: 3 May 2021

1. Introduction

The concepts of transformation optics (TO) [1,2] have been used to control the propagation characteristics of electromagnetic (EM) fields in interesting and useful ways by using regions of non-homogeneous, anisotropic materials. Based on the form-invariant nature of Maxwell's equations [3], the TO technique leads to implementation of unconventional electromagnetic devices using novel wave–matter interactions computed with coordinate transformations. One salient example is the cloak [4]. This success led to the development of many other unique EM devices [5–15] that exhibit unconventional and unusual propagation characteristics.

Phased array antennas have garnered significant interest in wireless communications applications due to their capability to change the shape and direction of the radiation pattern without physically moving the antennas. This technique is often referred to as beam-steering and can be accomplished by rotating the antenna elements or changing the relative phases of the radio-frequency (RF) signals driving the elements. Beam-steering of a phased array antenna is often a challenging task because it involves synthesis of multiple antenna elements and integration of control circuits, including solid-state phase shifters [16] and beam-forming networks [17], to control or guide the beam in a desired direction, as shown in Figure 1a. Recently, researchers showed that the TO technique could be useful to control the beam in a specific direction using coordinate transformations-based non-homogeneous material regions. Rahm et al. [7] showed how to design a beam shifter using TO.

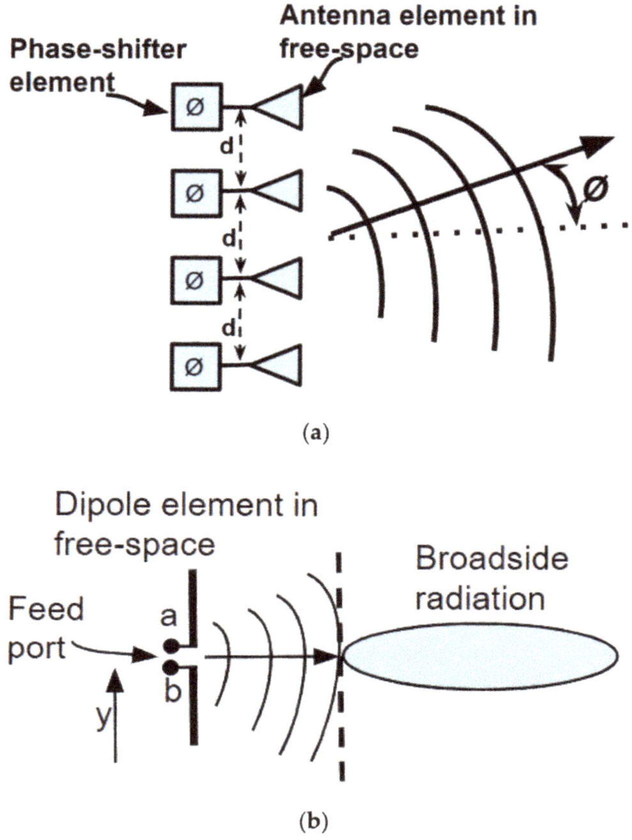

Figure 1. (**a**) A typical phased array antenna for beam-scanning. (**b**) A dipole antenna element along the y-direction in free-space.

Utilizing a similar idea, researchers in [12,13] proposed a set of beam-shifters to avoid obstacles in the beam path. In [18], researchers experimentally showed a TO-based lens for beam control at microwave frequencies. This pioneering research paved the way for beam-steering of antennas using the TO technique. The concepts of TO were later expanded to design unique antennas [19,20] and phased array antennas for different conformal applications [21–23]. In [24], it was shown that the techniques of TO can be utilized to manipulate EM fields and rotate them in a specified direction.

This approach can thus be used to control radiation characteristics of an antenna element (as shown in Figure 1b) or an antenna array in free-space and to rotate it in a desired direction, hence realizing a beam-steerer using TO-based media. This specific TO approach results in material properties that require active tuning to achieve beam-steering, but significant advancements and attention given to reconfigurable material properties, specifically in tunable constitutive parameters (permittivity and permeability), could in the future allow for practical implementation of novel beam-steering techniques [25–28]. Misra et al. [25] demonstrated electrically tunable permittivity in $BaTiO_3$ under DC bias conditions. In [26], researchers showed the influence of DC bias and temperature on the dielectric permittivity to achieve switchable dielectric permittivity in a semifluorinated azobenzene. Significant research has been done to control the permeability of materials.

In [27] researchers proposed microfluidic split-ring resonators inside a flexible elastomeric material to achieve reconfigurable effective permeability. Agarwal et al. [28] demonstrated the preparation of adaptive hybrid capsules with microgel/SiO_2 composite walls with tunable shell permeability.

Therefore, the objective of this work was to present a design and application of a TO-based cylindrical rotator for beam-steering, where a single dipole antenna element (as shown in Figure 2a) and an antenna *array* (as shown in Figure 2b) are enclosed by a TO-based non-homogeneous, anisotropic material shell designed using the transformation introduced in [24] (shown by the dotted ring in Figure 2). Through numerical simulations, beam- steering of the TO-based single element and the antenna array was demonstrated without using any phase control circuitry.

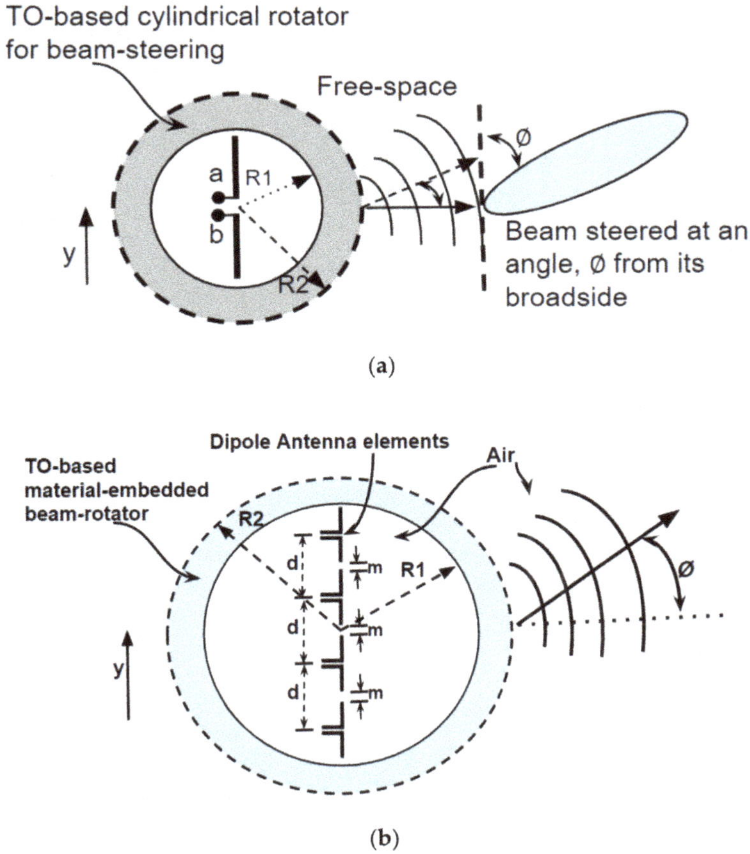

Figure 2. Metamaterial based cylindrical beam-steerer using TO: (**a**) Proposed TO-based cylindrical beam-steerer enclosing a single dipole element. (**b**) Material-embedded cylindrical beam-steerer using TO enclosing a co-linear vertical dipole array.

Then, the same TO-based cylindrical beam-rotator was applied to a vertical dipole antenna in free-space to design a horizontal dipole antenna. It was shown that the horizontal dipole element embedded in a metamaterial radiated in a similar manner as that of the vertical dipole element in free-space. Similarly, the TO-based cylindrical beam-rotator was also applied to a co-linear vertical array in free-space to design a horizontal co-linear array, where through numerical simulations it was shown that the material-embedded horizontal array radiates in the same manner as the vertical array in free-space. Finite element method-based full-wave simulations via COMSOL Multiphysics ® were used to numerically analyze and demonstrate the performance of the proposed TO-based beam-scanning technique. It should be noted that the theory that was validated by COMSOL Multiphysics is similar to the approaches taken by previous works and reported in [19–24].

2. Proposed TO-Based Single Element Cylindrical Beam-Steerer

2.1. Theoretical Model

Consider the dipole element positioned in free-space along the y-axis represented in Figure 1b. The dipole is of λ length, where λ is the free-space wavelength at which the dipole antenna is designed to operate. The current distribution of the dipole in Figure 1b along $x = 0$ was chosen to be [19,20]:

$$\bar{j} = \begin{bmatrix} 0 \\ \frac{1}{\sqrt{\sigma * \pi}} \cdot e^{-\frac{y^2}{\sigma}} \\ 0 \end{bmatrix} \tag{1}$$

where σ is much smaller than the length of the dipole. The current distribution model is a way of handling sheet current as the limit of a volumetric current density, which was suggested in [19,20]. The sigma parameter is set to be infinitesimally small relative to the length of the dipole and the limit $\sigma \to 0$ can be taken to approximate the current distribution on a thin wire at $x = 0$ [19,20]. The intent is to introduce a TO- based material shell (again as illustrated as the grey ring in Figure 2a) to control the radiation characteristics of the dipole element to steer its beam to a desired direction. Starting with the basic transformation media approach, the associated permittivity and permeability tensors of transformation media are given by [29]:

$$\varepsilon' = \mu' = \frac{A \varepsilon A^T}{det A} \tag{2}$$

where $A = \partial(x', y', z')/\partial(x,\ y,\ z)$ is the Jacobian matrix and A^T is the transpose of the Jacobian. The mapping between the original $(r,\ \theta,\ z)$ and the transformed (r', θ', z') cylindrical coordinate systems are [24]:

$$r' = r, \tag{3}$$

$$\theta' = \begin{cases} \theta + \beta,\ r < R_1 \\ \theta + \frac{\beta(R_2 - r)}{R_2 - R_1},\ R_1 \le r < R_2 \\ \theta,\ r \ge R_2 \end{cases} \tag{4}$$

$$z' = z, \tag{5}$$

where β is the angle of rotation in the region between radii R_1 and R_2 in Figure 2. By controlling the rotation angle β, it is possible to control the radiation characteristics and by extension the amount of beam-steering in a desired direction using a single antenna element and the array without phase control circuitry and multiple antenna elements, as shown in Figure 2. The transformation Equation (2) yields the permittivity and permeability tensors of the material between $r = R_1$ and $r = R_2$ as [24]:

$$\varepsilon' = \mu' = \begin{pmatrix} 1 + 2df + d^2 sin^2\theta & -d^2 f - dg & 0 \\ -d^2 f - dg & 1 - 2df + d^2 cos^2\theta & 0 \\ 0 & 0 & 1 \end{pmatrix}, \tag{6}$$

where $d = \frac{\beta * r}{R_2 - R_1}$, $f = cos\theta sin\theta$, $g = cos^2\theta - sin^2\theta$, and $\varepsilon' = \mu' = I$ in other regions in Figure 2.

Equation (6) results in anisotropic and inhomogeneous permittivity and permeability tensors for the spherical shell. Note that the results lead to a perfect impedance matching with no reflection at the boundaries of the material region and free-space, which is shown in Figure 3. Such an anisotropic and inhomogeneous transformation medium can be realized by discrete metamaterials and structures such as periodic split ring resonators (SRRs) [30]. The theoretical material parameters in Equation (6) were validated using full-wave simulations in the finite element solver COMSOL Multiphysics ®, as shown in Figure 3. An incident TE plane wave of 10 GHz frequency was used from left to right along the x-direction. The inner radius $R_1 = 1.1\lambda$, and the outer radius $R_2 = 2\lambda$. The radii R_1 and R_2 of the metamaterial coating were chosen by closely following the similar works reported in [24] and [31]. As the dipole element was full-wave ($L = \lambda$), it was necessary to choose an inner radius R_1 of the metamaterial coating that was bigger than the length of the dipole, so that it follows the transformation rule from Equation (4). The metamaterial coating is located in the radiative near field region, as $0.62\sqrt{\frac{D^3}{\lambda}}$ < metamaterial coating < $\frac{2D^2}{\lambda}$, where D = L = maximum linear dimension of the antenna, and λ = wavelength of the EM wave.

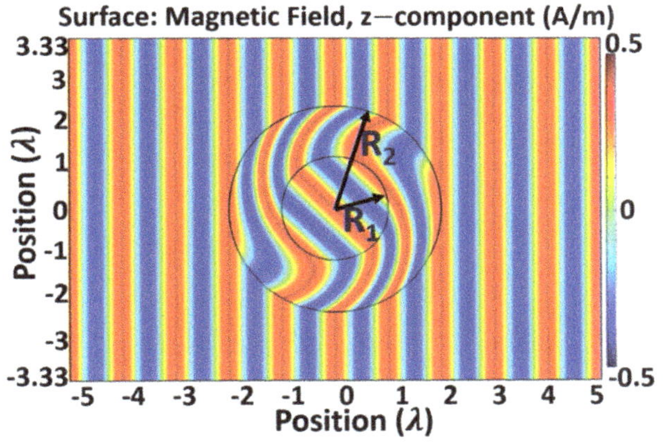

(a)

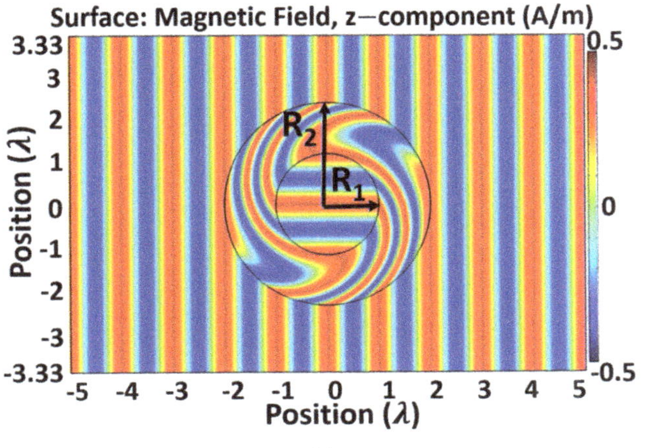

(b)

Figure 3. Verification of material parameters for TO-based cylindrical rotator showing perfect TEM wave with no scattering at the boundaries of the material region and free-space (**a**) at rotation angle $\beta = 45°$, (**b**) at rotation angle $\beta = 90°$.

The material parameters were calculated using Equation (6) for different angles of rotation (β). Numerical simulations were also completed to see the spatial variation of the constitutive material parameters, as shown in Figure 4, and it was observed that the material parameters were well within the range of material properties (permittivity and permeability) mentioned in [25–27]. The rotation angle β was chosen to be 45° for the simulations.

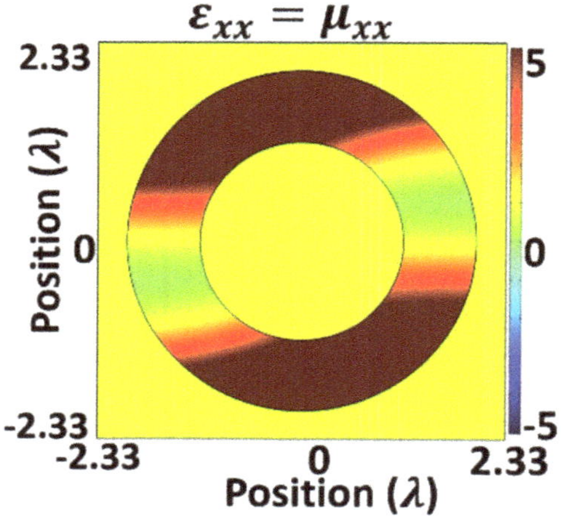

(a)

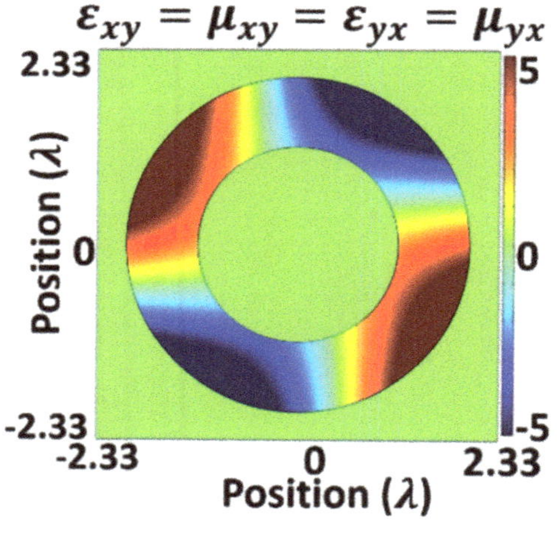

(b)

Figure 4. *Cont.*

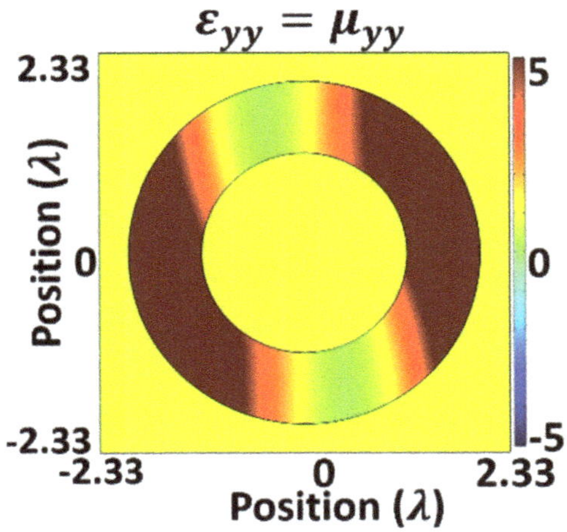

(c)

Figure 4. Spatial variation of material parameters inside the TO-based rotator shell, (**a**) xx-components, (**b**) xy- and yx- components, and (**c**) yy-components. The dimensions of the rotator are given by $R_1 = 1.17\lambda$ and $R_2 = 2\lambda$. The material parameters ε_{xy}, μ_{xy}, ε_{yx}, and μ_{yx} are equal.

Next, a full-wave (λ) dipole antenna element was placed in the region $r < R_1$ and the material parameters from Equation (6) were used to design the cylindrical beam-steerer in the region $R_1 \leq r < R_2$ to control the radiation characteristics of the dipole element in a desired direction, as shown in Figure 2a. The beam-steering angle ϕ of the dipole antenna element is controlled by the rotation angle β. Equations (4) and (6) show that it is possible to rotate the EM fields in an arbitrary direction, which makes the beam-rotator capable of steering the beam.

2.2. Full-Wave Simulation Results

The objective of this research was to exploit the concepts of transformation electromagnetics/optics (TE/TO) for realizing a beam-steering technique using a rotation mapping introduced in [24]. For simplicity and ease of coordinate transformation, here, no transformation was considered along the z-direction. As a result, a 2D transformation media was chosen, which resulted in material parameters in Equation (6). An experimental realization of the rotation coating requires building blocks that have anisotropic dielectric functions, and a similar theory of this kind of rotation mapping could be extended to 3D. In that case, we will have permittivity and permeability tensors in Equation (6) due to variations along the z-direction.

The performance of the proposed single element TO-based beam-steerer, as shown in Figure 2a, was demonstrated through numerical solutions in the finite-element simulation software COMSOL Multi-physics ®. Figure 5 presents the y-component of the electric field of the proposed single element beam-rotator verifying the transformed media from (6). Figure 5a shows the simulation results from a full-wave ($L = \lambda$) dipole antenna in free-space along the y-direction (Figure 1b). This will be called the "vertical dipole". A frequency of 10 GHz was chosen. Now, to control the radiation characteristics of the dipole element in a desired direction, the transformation media from (6) was used as the beam-steerer around the dipole and the rotation angle β in (4) was controlled to steer the beam of the dipole antenna element in the desired direction. A rotation angle $\beta = 22.5°$ was chosen to rotate the fields pattern of the vertical dipole at an angle of 22.5°, as a result, the beam was steered at an angle $\phi = 22.5°$, as shown in Figure 5b. Similarly, rotation angle $\beta = 45°$ was chosen to steer the beam of the vertical dipole to an angle $\phi = 45°$ (Figure 5c). Figure 5d shows electric field radiation of a full-wave ($L = \lambda$) dipole antenna in free-space along the x-direction. This is denoted as the "horizontal dipole" antenna. The current distribution from (1) was re-defined for the horizontal dipole as the location of the dipole changed to y = 0 from x = 0. Now, the TO-based beam-rotator was used and a rotation angle $\beta = 90°$ was chosen to transform the horizontal dipole into the vertical dipole, as shown in Figure 5e. The fields from the vertical dipole in Figure 5a and the transformed horizontal dipole antenna in Figure 5e outside the material shell are the same. Figure 5f verifies that the difference between the two fields is negligible and there is almost no field distribution outside the transformation media.

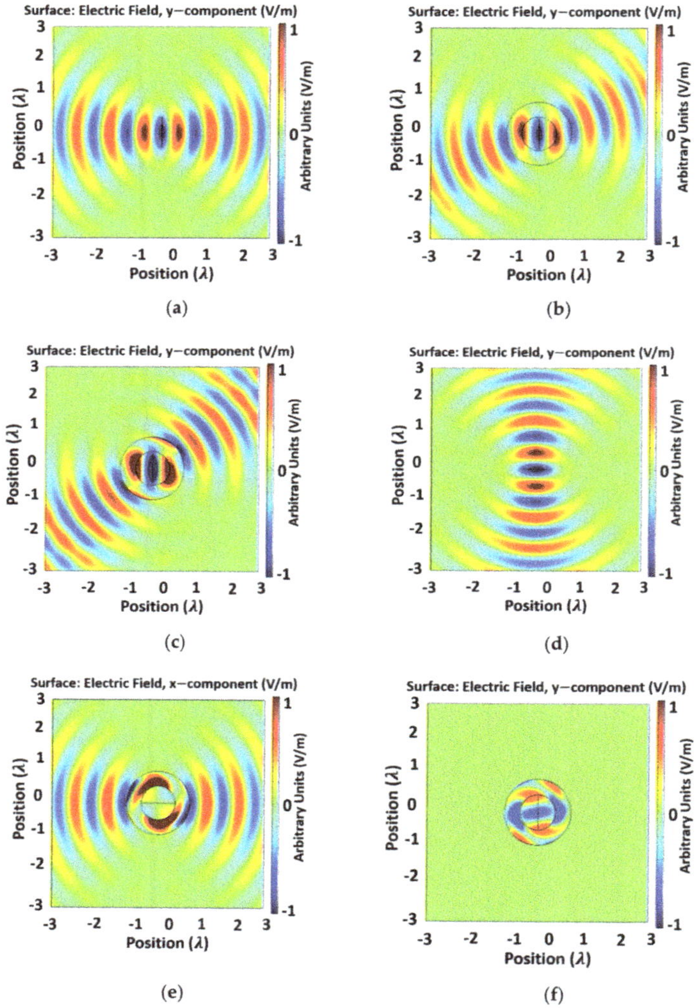

Figure 5. The electric fields of the proposed TO-based single element beam-steerer (**a**) dipole antenna of length $L = \lambda$ in free-space along y-direction (vertical); (**b**) the fields of the vertical dipole that have undergone a rotation of $\beta = 22.5°$; (**c**) the fields of the vertical dipole that have undergone a rotation of $\beta = 45°$; (**d**) dipole antenna of length $L = \lambda$ in free-space along x-direction (horizontal); (**e**) the fields of the horizontal dipole that have undergone a rotation of $\beta = 90°$; (**f**) difference between the magnetic fields in (**a**) and (**e**).

Furthermore, the far-field radiation patterns of the TO-based beam-rotator using a single antenna element were simulated and are illustrated in Figure 6. As shown in Figure 6a, the transformation media from (6) was used to rotate the beam of the vertical dipole in free-space to angles of $\phi = 22.5°$ and $\phi = 45°$ by setting the rotation angles $\beta = 22.5°$ and $\beta = 45°$ in transformation media from (6), respectively. Moreover, Figure 6b shows that the radiation pattern of the vertical dipole in free-space is similar to the radiation pattern of the horizontal dipole, when the horizontal dipole is enclosed by the transformation media and is rotated by $\beta = 90°$, but is different if the horizontal dipole is not enclosed by the transformed medium and is not rotated by an angle $\beta = 90°$.

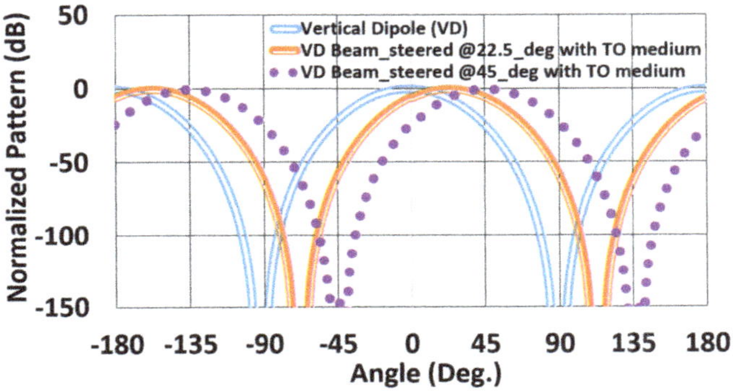

(a)

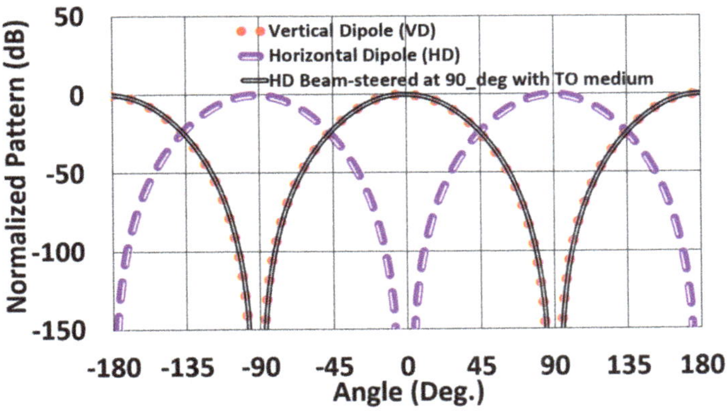

(b)

Figure 6. Far-field radiation pattern of proposed TO-based single element (**a**) beam-scheme 22.5° and $\phi = 45°$; (**b**) beam-steering of the horizontal dipole at $\phi = 90°$.

Moreover, considering that losses exist in practical materials, numerical simulations were performed incorporating different values of loss tangents ($\tan\delta$), as shown in Figure 7. Loss was incorporated in the simulations by replacing ε_{xx} with $(\varepsilon_{xx} - j|\varepsilon_{xx}|\tan\,\delta)$ [22,23]. Other tensor parameters were also modified similarly. A rotation angle $\beta = 22.5°$ was chosen to steer the beam at 22.5°. Figure 7 shows that while the antenna's radiated field strength degrades with the increase of loss factor, its overall steering capability remains unchanged. With the loss tangent reduced to only 0.1, the effect of loss is almost unnoticeable. No significant differences were observed in the range $\tan\delta \leq 0.01$.

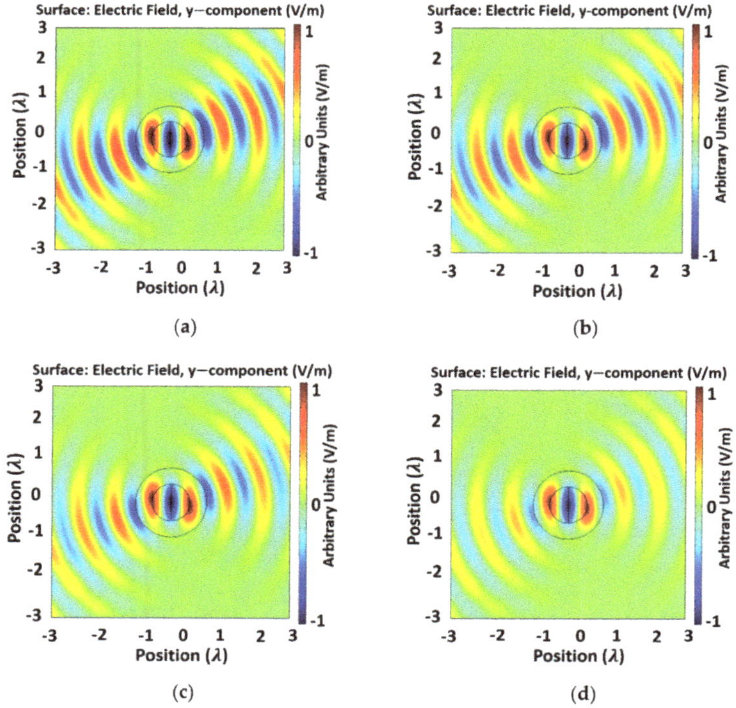

Figure 7. The electric fields for the proposed TO-based single element beam-steerer for different values of loss factor ($\tan\delta$): (**a**) $\tan\delta = 0.0$; (**b**) $\tan\delta = 0.01$; (**c**) $\tan\delta = 0.1$; (**d**) $\tan\delta = 0.3$.

3. Proposed Antenna Array Enclosed by TO-Based Cylindrical Beam-Steerer

The same TO technique from Section 2.1 can be utilized to realize a cylindrical beam-rotator enclosing an antenna array with the TO-based non-homogeneous, anisotropic media. Next, consider the N-element co-linear dipole array along the y-axis represented in Figure 2b. Each of the elements in the array are equally spaced with the edge-to-edge distance between the elements of $\mathrm{m} = \lambda/15$, where λ is the free-space wavelength at which the phased array is designed to operate. A two-dimensional (2D) space is considered to illustrate the proposed array. The current distribution of each of the dipole elements in the array is approximated as the current distribution of a thin wire along $x = 0$ and is defined by Equation (1). In this case, an array of four elements is chosen to validate the proposed beam scanning method. Each of the dipoles in the array is of $\lambda/2$ length spanning over a distance of 2.2λ. There was no phase difference considered between the adjacent dipole elements. Here, a TO-based material shell enclosing the dipole array to control the radiation characteristics of the array and steer its beam to a desired direction as shown in Figure 2b is introduced.

Next, the four-element dipole antenna array was placed in the region $r < R_1$ and the material parameters from Equation (6) were used to design the cylindrical beam-steerer in the region $R_1 \le r < R_2$ enclosing the array to control the radiation characteristics of the antenna array in a desired direction, as shown in Figure 2b. The beam-scanning angle $\varnothing_s$ of the dipole antenna array will be controlled by the rotation angle β from (6). From Equations (4) and (6), it is shown that it is possible to rotate the EM fields in an arbitrary direction, which makes the beam-rotator capable of steering the dipole antenna array pattern in a desired direction, thus enabling antenna array scanning.

Full-Wave Simulations Results

The performance of the proposed phased array antenna enclosed by TO-based material-embedded cylindrical beam-steerer, as shown in Figure 2b, was demonstrated through numerical solutions in the commercially available finite-element simulation software COMSOL Multiphysics ®. Figure 8 presents the y-component of the electric field of the proposed beam-rotator for scanning of the phased array antenna verifying the transformed media from (6). Figure 8a demonstrates the simulation results from the dipole antenna array in free-space along the y-direction (as shown in Figure 2b). For reference, it will be called the "vertical array". A frequency of 10 GHz was chosen. To control the radiation characteristics of the dipole antenna array in a desired direction, the transformation media from (6) was used as the beam-steerer around the array and the rotation angle β in (4) was controlled to steer the beam of the dipole antenna array in the desired direction. A rotation angle $\beta = 22.5°$ was chosen to rotate the field patterns of the "vertical array" at an angle of 22.5°, as a result, a beam-scanning of the "vertical dipole array" occurred at an angle $\varnothing_s = 22.5°$, as shown in Figure 8b. Similarly, rotation angle $\beta = 45°$ was chosen to scan the "vertical dipole array" beam to an angle $\varnothing_s = 45°$ (as demonstrated in Figure 8c).

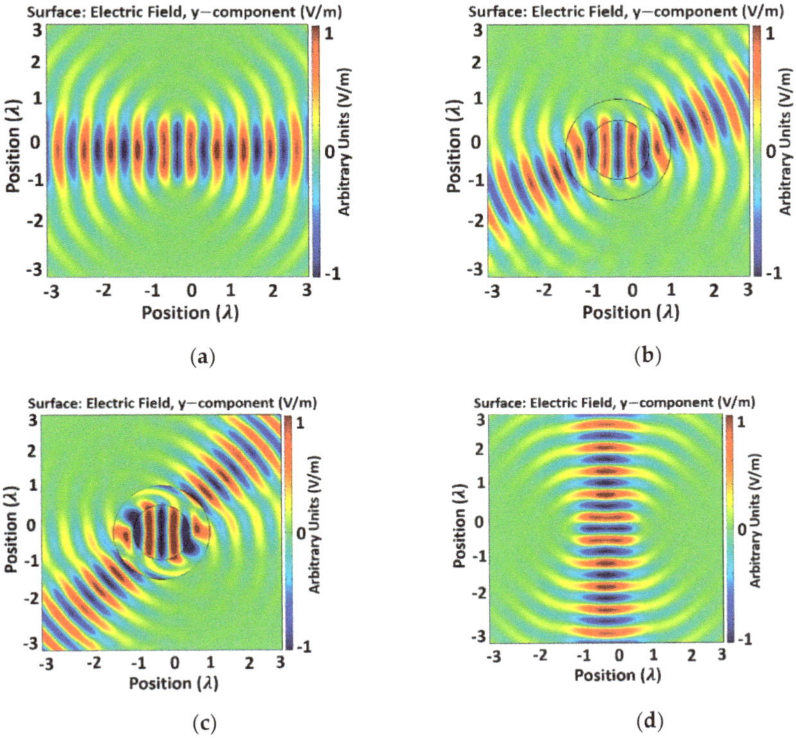

Figure 8. *Cont.*

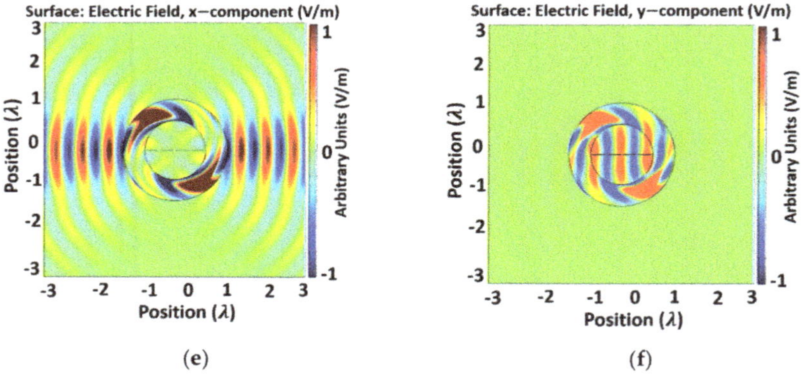

Figure 8. The electric fields of the proposed array antenna enclosed by TO-based material-embedded cylindrical beam-rotator for beam-scanning (**a**) dipole antenna array in free-scheme, (**b**) The fields of the vertical dipole array enclosed by TO-based material shell and scanned at $\varnothing_s = 22.5°$, (**c**) The fields of the vertical dipole array enclosed by TO-based material shell and scanned at $\varnothing_s = 45°$; (**d**) dipole antenna array in free-space along the x-direction (horizontal array); (**e**) the fields of the horizontal dipole array enclosed by a TO-based material shell and that has undergone a rotation of $\varnothing_s = 90°$; (**f**) difference between the electric fields in (**a**) and (**e**).

Figure 8d presents the electric field radiation of a dipole antenna array in free-space along the x-direction. For reference, it is denoted as the "horizontal array". The current distribution from (1) was adjusted for each of the elements of the "horizontal array" as the location of each dipole element changed to $y = 0$ from $x = 0$. Now, the "horizontal array" was enclosed by the proposed TO-based beam-rotator and a rotation angle $\beta = 90°$ was chosen to transform the "horizontal dipole array" into the "vertical dipole array", as shown in Figure 8e. The fields from the "vertical dipole array" in Figure 8a and the transformed "horizontal dipole array" in Figure 8e outside the material shell are the same. This is emphasized in Figure 8f, which shows almost no field distribution outside the transformation media when the difference between the two fields is taken, validating the results further.

Moreover, the far-field patterns of the proposed array enclosed by the TO-based material-embedded cylindrical beam-rotator were simulated and are illustrated in Figure 9. As shown in Figure 9a, the transformation media from (6) was used to rotate the beam of the "vertical array" in free-space to angles $\varnothing_s$= 22.5° and $\varnothing_s$= 45° by setting the rotation angles $\beta = 22.5°$ and $\beta = 45°$ in transformation media from (6), respectively. Moreover, Figure 9b shows that the radiation pattern of the "vertical array" in free-space is similar to the radiation pattern of the "horizontal array", when the "horizontal array" is enclosed by the transformation media and is rotated by $\beta = 90°$, but is different if the "horizontal array" is not enclosed by the transformed medium and is not rotated by an angle $\beta = 90°$.

Furthermore, since practical metamaterial designs have losses, finite-element full-wave simulations were performed adding different values of loss tangent ($\tan\delta$). For a scan angle $\varnothing_s = 90°$, the normalized radiation patterns of the TO-based "horizontal array" for different values of loss tangent ($\tan\delta$) are compared in Figure 10. Loss was incorporated in the simulations by replacing ε_{xx} with $(\varepsilon_{xx} - j|\varepsilon_{xx}|tan\ \delta)$ [22,23]. Similar modifications were made in other tensor material parameters. As loss is increased, the antenna's radiated field strength degrades, but its overall steering capability remains unchanged, which is shown in Figure 10. With the loss tangent reduced to only 0.1, the effect of loss is almost insignificant. No noticeable differences were observed in the range $\tan\delta \leq 0.01$.

Dispersion exists in all materials and systems, natural or manmade. There are many ways to implement the needed material properties given the frequency regime, application, environmental considerations, etc. Dispersion along with other fundamental properties such a loss, noise, etc. must always be taken into consideration to meet a specific application system's requirements. Several research works [32–34] have been performed to explore and analyze the limitations of specific implementations of TO devices due to dispersive materials. To this end, we proposed a TO-based beam-steering technique that results in anisotropic, non-homogeneous material parameters. Keeping the practical implementation of metamaterials and its dispersive nature in mind, numerical simulations are presented in Figures 7 and 10 by incorporating losses to see how the losses in the material parameters in Equation (6) affect the performances of the proposed beam-rotator. It is anticipated that the material parameters from Equation (6) will demonstrate a dispersive nature when practically implemented.

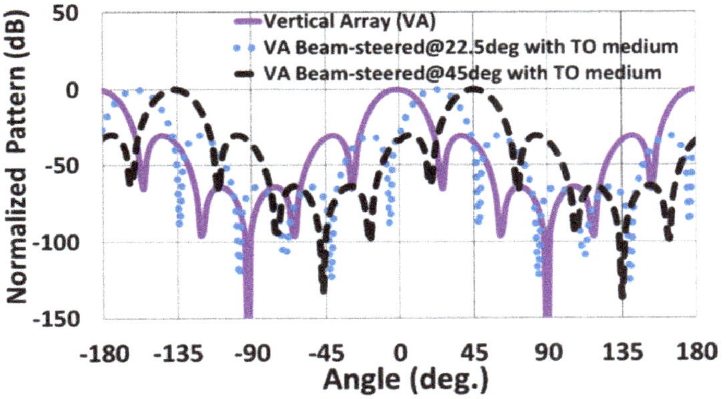

(**a**)

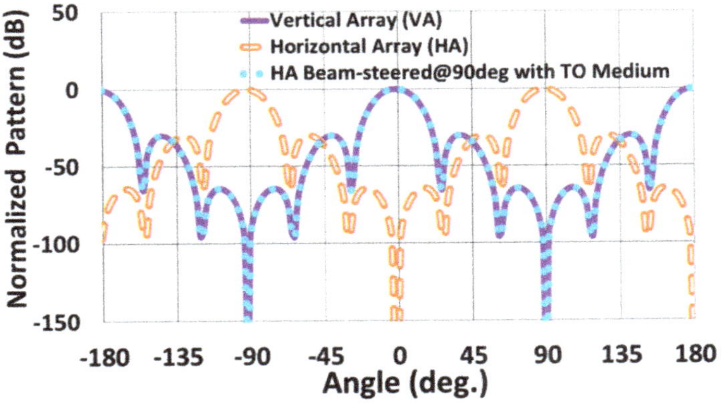

(**b**)

Figure 9. Far-field radiation pattern of proposed antenna array enclosed by TO-based material-embedded cylindrical beam-rotator (**a**) beam-scanning of the "virtual array" at $\varnothing_s = 22.5°$ and $\varnothing_s = 45°$; (**b**) beam-scanning of the "horizontal array" at $\varnothing_s = 90°$.

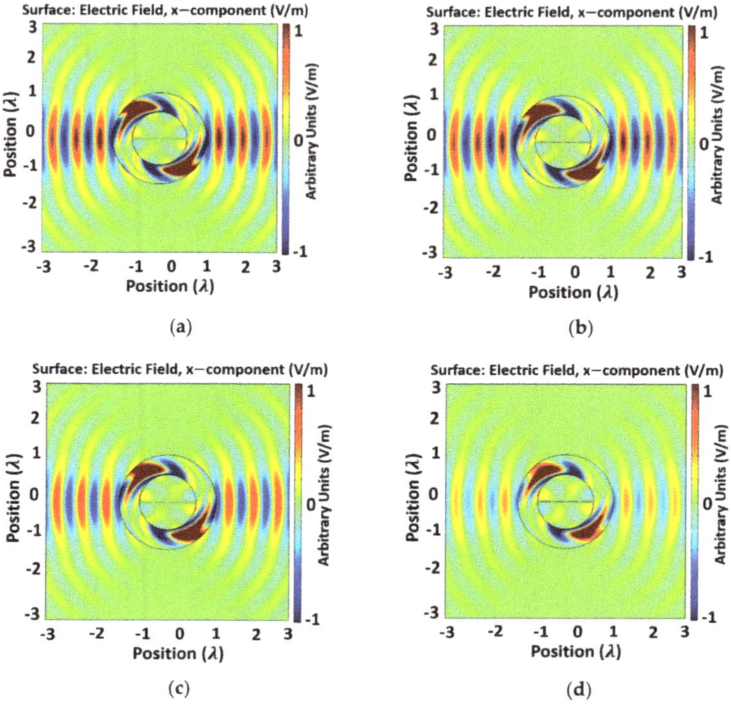

Figure 10. The electric fields for the proposed TO-based "horizontal array" for different values of loss factor ($\tan\delta$): (**a**) $\tan\delta = 0.0$; (**b**) $\tan\delta = 0.01$; (**c**) $\tan\delta = 0.1$; (**d**) $\tan\delta = 0.3$.

4. Conclusions

In conclusion, it has been shown how transformation optics can be utilized to steer a beam in an arbitrary direction from a single antenna element and an antenna array without using phase control circuitry and beam-forming networks. The proposed beam-steerer is a TO-based non-homogeneous, anisotropic material shell theoretically computed using coordinate transformations. The transformed parameters are derived, and through full-wave simulations, the beam-steering performances of the TO-based beam-steerer are demonstrated. Additionally, the TO-based beam-rotator is applied to the vertical dipole antenna in free-space to design a horizontal dipole antenna, and through numerical simulations it is shown that the material-embedded horizontal dipole element radiates as a vertical dipole element in free-space, verifying the design. Similarly, the TO-based beam-rotator is applied to a vertical dipole array to design a TO-based material-embedded horizontal dipole array, which behaves like a vertical dipole array in free-space. While we have presented numerical verification, in practice, this TO approach requires actively tuned material parameters. To this end, significant advancements have been made to realize actively tunable constitutive material parameters, which could enable practical implementation of this TO-based beam-steering technique.

Author Contributions: Conceptualization, D.M.; methodology, D.M.; software, D.M.; validation, D.M., S.D. and J.W.A.; formal analysis, D.M.; investigation, D.M., B.D.B., S.D. and J.W.A.; resources, B.D.B.; writing—original draft preparation, D.M.; writing—review and editing, D.M., S.D., J.W.A., M.S.A. and B.D.B.; supervision, B.D.B., M.S.A., and J.W.A.; project administration, B.D.B., M.S.A. and J.W.A.; funding acquisition, B.D.B., M.S.A. and J.W.A. All authors have read and agreed to the published version of the manuscript.

Funding: This research was funded by US Air Force Research Laboratory Munitions Directorate under Grant DSC6511-01.

Data Availability Statement: Data is available upon reasonable request from the corresponding authors.

Conflicts of Interest: The authors declare no conflict of interest.

References

1. Pendry, J.B.; Schurig, D.; Smith, D.R. Controlling Electromagnetic Fields. *Science* **2006**, *312*, 1780–1782. [CrossRef]
2. Leonhardt, U. Optical Conformal Mapping. *Science* **2006**, *312*, 1777–1780. [CrossRef]
3. Post, E.J. *Formal Structure of Electromagnetics*, 1st ed.; Dover Publications, Inc.: New York, NY, USA, 1962.
4. Schurig, D.; Mock, J.J.; Justice, B.J.; Cummer, S.A.; Pendry, J.B.; Starr, A.F.; Smith, D.R. Metamaterial electromagnetic cloak at microwave frequencies. *Science* **2006**, *314*, 977–980. [CrossRef]
5. Chen, H.; Luo, X.; Ma, H.; Chan, C.T. The anti-cloak. *Opt. Express* **2008**, *16*, 14603–14608. [CrossRef]
6. Yang, T.; Chen, H.; Luo, X.; Ma, H. Superscatterer: Enhancement of scattering with complementary media. *Opt. Express* **2008**, *16*, 18545–18550. [CrossRef] [PubMed]
7. Rahm, M.; Cummer, S.A.; Schurig, D.; Pendry, J.B.; Smith, D.R. Optical design of reflectionless complex media by finite embedded coordinate transformations. *Phys. Rev. Lett.* **2008**, *100*, 063903. [CrossRef]
8. Yi, J.; Tichit, P.; Burokur, S.N.; de Lustrac, A. Illusion optics: Optically transforming the nature and the location of electromagnetic emissions. *J. Appl. Phys.* **2015**, *117*, 084903. [CrossRef]
9. Tichit, P.; Burokur, S.N.; Lustrac, A.D. Ultradirective antenna via transformation optics. *J. Appl. Phys.* **2009**, *105*, 104912. [CrossRef]
10. Tichit, P.; Burokur, S.N.; Lustrac, A.D. Transformation media producing quasi-perfect isotropic emission. *Opt. Express* **2011**, *19*, 20551–20556. [CrossRef] [PubMed]
11. Kwon, D.H.; Werner, D.H. Transformation optical designs for wave collimators, flat lenses and right-angle bends. *New J. Phys.* **2008**, *10*, 115023. [CrossRef]
12. Wang, M.Y.; Zhang, J.J.; Chen, H.; Luo, Y.; Xi, S.; Ran, L.-X.; Kong, J. Design and Application of a Beam Shifter by Transformation Media. *Prog. Electromagn. Res.* **2008**, *83*, 147–155. [CrossRef]
13. Mitra, D.; Cleveland, J.; Lewis, J.; Braaten, B.D.; Allen, J.; Allen, M. On the use of Multiple Beam-Shifters Developed using Transformation Electromagnetics to Control Propagation Characteristics. In Proceedings of the IEEE Research and Applications of Photonics in Defense Conference (RAPID), Miramar Beach, FL, USA, 10–12 August 2020; pp. 1–2.
14. Rahm, M.; Roberts, D.A.; Pendry, J.B.; Smith, D.R. Transformation-optical design of adaptive beam bends and beam expanders. *Opt. Express* **2008**, *16*, 11555–11567. [CrossRef] [PubMed]
15. Tichit, P.; Burokur, S.N.; Yi, J.; de Lustrac, A. Transformation Electromagnetics for Antennas with an Illusion on the Radiation Pattern. *IEEE Antennas Wirel. Propag. Lett.* **2014**, *13*, 1796–1799. [CrossRef]
16. Mitra, D.; Roy, P.; Dawn, D. A variable gain CMOS phase shifter for phased array antenna applications. *Microw. Opt. Technol. Lett.* **2016**, *59*, 324–328. [CrossRef]
17. Sanford, J.R. Design of a Miniature Reactive Beam Forming Network. In Proceedings of the 2019 IEEE International Symposium on Antennas and Propagation and USNC-URSI Radio Science Meeting, Atlanta, GA, USA, 7–12 July 2019; pp. 1357–1358.
18. Yi, J.; Burokur, S.N.; Lustrac, A.D. Experimental validation of a transformation optics based lens for beam steering. *Appl. Phys. Lett.* **2015**, *107*, 154101. [CrossRef]
19. Kundtz, N.; Roberts, D.A.; Allen, J.; Cummer, S.; Smith, D.R. Optical source transformations. *Opt. Express* **2008**, *16*, 21215–21222. [CrossRef] [PubMed]
20. Allen, J.; Kundtz, N.; Roberts, D.A.; Cummer, S.A.; Smith, D.R. Electromagnetic source transformations using superellipse equations. *Appl. Phys. Lett.* **2009**, *94*, 194101. [CrossRef]
21. Popa, B.; Allen, J.; Cummer, S.A. Conformal Array Design with Transformation Electromagnetics. *Appl. Phys. Lett.* **2009**, *94*, 244102. [CrossRef]
22. Kwon, D.H. Virtual Circular Array using Material-Embedded Linear Source Distributions. *Appl. Phys. Lett.* **2009**, *95*, 173503. [CrossRef]
23. Deng, L.; Wu, Y.; Hong, W.; Zhu, J.; Peng, B.; Li, S. Conformal Array Design on Arbitrary Polygon Surface with Transformation Optics. *AIP Adv.* **2016**, *6*, 065011. [CrossRef]
24. Chen, H.; Chan, C.T. Transformation media that rotate electromagnetic fields. *Appl. Phys. Lett.* **2007**, *90*, 241105. [CrossRef]
25. Misra, S.; Kalaswad, M.; Zhang, D.; Wang, H. Dynamic tuning of dielectric permittivity in $BaTiO_3$ via electrical biasing. *Mater. Res. Lett.* **2020**, *8*, 321–327. [CrossRef]
26. Stangenberg, R.; Grigoriadis, C.; Butt, H.; Müllen, K.; Floudas, G. Switchable permittivity with temperature and DC-bias in a semifluorinated azobenzene derivative. *Colloid Polym. Sci.* **2014**, *292*, 1939–1948. [CrossRef]
27. Kasirga, T.S.; Ertas, Y.N.; Bayindir, M. Microfluidics for reconfigurable electromagnetic metamaterials. *Appl. Phys. Lett.* **2009**, *95*, 214102. [CrossRef]
28. Agarwal, G.; Ulpenich, A.; Zhu, X.; Möller, M.; Pich, A. Microgel-Based Adaptive Hybrid Capsules with Tunable Shell Permeability. *Chem. Mater.* **2014**, *26*, 5882–5891. [CrossRef]

29. Schurig, D.; Pendry, J.B.; Smith, D.R. Calculation of material properties and ray tracing in transformation media. *Opt. Express* **2006**, *14*, 9794–9804. [CrossRef] [PubMed]
30. Pendry, J.B.; Holden, A.J.; Robbins, D.J.; Stewart, W.J. Magnetism from conductors and enhanced nonlinear phenomena. *IEEE Trans. on Microw. Theory Technol.* **1999**, *47*, 2075–2084. [CrossRef]
31. Kwon, D.H.; Emiroglu, C.D. Low-profile embedded design of endfire scanning arrays with coordinate transformation. *J. Appl. Phys.* **2010**, *107*, 034508. [CrossRef]
32. Yao, P.; Liang, Z.; Jiang, X. Limitation of the electromagnetic cloak with dispersive material. *Appl. Phys. Lett.* **2008**, *92*, 031111. [CrossRef]
33. Hashemi, H.; Qiu, C.-W.; McCauley, A.P.; Joannopoulos, J.D.; Johnson, S.G. Diameter-bandwidth product limitation of isolatedobject cloaking. *Phys. Rev. A* **2012**, *86*, 013804. [CrossRef]
34. Cassier, M.; Milton, G.W. Bounds on Herglotz functions and fundamental limits of broadband passive quasi-static cloaking. *J. Math. Phys.* **2017**, *58*, 071504. [CrossRef]

MDPI
St. Alban-Anlage 66
4052 Basel
Switzerland
Tel. +41 61 683 77 34
Fax +41 61 302 89 18
www.mdpi.com

Electronics Editorial Office
E-mail: electronics@mdpi.com
www.mdpi.com/journal/electronics

www.ingramcontent.com/pod-product-compliance
Lightning Source LLC
Chambersburg PA
CBHW061627210926
51850CB00009B/2457
* 9 7 8 3 0 3 6 5 8 0 3 5 7 *